High-Level Test Synthesis of Digital VLSI Circuits

The Artech House Solid-State Technology Library

Advanced Semiconductor Device Physics and Modeling, Juin J. Liou

Amorphous and Microcrystalline Semiconductor Devices, Volume II: Materials and Device Physics, Jerzy Kanicki, editor

Computational Modeling in Semiconductor Processing, M. Meyyappan, editor

Digital Hardware Testing: Transistor-Level Fault Modeling and Testing, Rochit Rajsuman

Electrical and Thermal Characterization of MESFETs, HEMTs, and HBTs, Robert Anholt

High-Level Test Synthesis of Digital VLSI Circuits, Mike Tien-Chien Lee

Iddq Testing for CMOS VLSI, Rochit Rajsuman

Indium Phosphide and Related Materials: Processing, Technology, and Devices, Avishay Katz

InP HBTs: Growth, Processing, and Applications, B. Jalali and S. J. Pearton, editors

Introduction to Semiconductor Device Yield Modeling, Albert V. Ferris-Prabhu

Microelectronic Reliability, Volume II: Integrity Assessment and Assurance, Emiliano Pollino

Modern GaAs Processing Methods, Ralph E. Williams

Numerical Simulation of Submicron Semiconductor Devices, Kazutaka Tomizawa

Principles and Analysis of AlGaAs/GaAs Heterojunction Bipolar Transistors, Juin J. Liou

Reliability and Degradation of III-V Optical Devices, Osamu Ueda

Systematic Analysis of Bipolar and MOS Transistors, Ugur Cilingiroglu

Vacuum Mechatronics, Gerardo Beni et al

VLSI Metallization: Physics and Technologies, Krishna Shenai

For further information on these and other Artech House titles, contact:

Artech House
685 Canton Street
Norwood, MA 02062
617-769-9750
Fax: 617-769-6334
Telex: 951-659
email: artech@artech-house.com
WWW: http://www.artech-house.com

Artech House
Portland House, Stag Place
London SW1E 5XA England
+44 (0) 171-973-8077
Fax: +44 (0) 171-630-0166
Telex: 951-659
email: artech-uk@artech-house.com

High-Level Test Synthesis of Digital VLSI Circuits

Mike Tien-Chien Lee

Artech House
Boston • London

Library of Congress Cataloging-in-Publication Data
Lee, Mike Tien-Chien
    High-level test synthesis of digital VLSI circuits / Mike Tien-Chien Lee.
       p.   cm.
    Includes bibliographical references and index.
    ISBN 0-89006-907-7 (alk. paper)
      1.Integrated circuits—Very large scale integration—Design and construction—Data processing.   2. Integrated circuits—Very large scale integration—Testing—Data processing.   3. Computer-aided design.
4. Digital integrated circuits—Testing—Data processing.
    I. Title.
  TK7874.75.L44   1997
  621.39'5—dc21   96-6584
                                                                         CIP

British Library Cataloguing in Publication Data
Lee, Mike Tien-Chien
    High-level synthesis of digital VLSI circuits
    1. Integrated circuits—Very large scale integration 2. Integrated circuits—Verification
    I. Title
    621.3'95
    ISBN 0-89006-907-7
Cover design by Jennifer Makower

© 1997 ARTECH HOUSE, INC.
685 Canton Street
Norwood, MA 02062

All rights reserved. Printed and bound in the United States of America. No part of this book may be reproduced or utilized in any form or by any means, electronic or mechanical, including photocopying, recording, or by any information storage and retrieval system, without permission in writing from the publisher.

    All terms mentioned in this book that are known to be trademarks or service marks have been appropriately capitalized. Artech House cannot attest to the accuracy of this information. Use of a term in this book should not be regarded as affecting the validity of any trademark or service mark.

International Standard Book Number: 0-89006-907-7
Library of Congress Catalog Card Number: 96-6584
10 9 8 7 6 5 4 3 2 1

Contents

Preface   ix

1 Introduction   1
    1.1 Circuit Testing 3
    1.2 High-Level Synthesis in VLSI Design 7
    1.3 High-Level Test Synthesis 13
    1.4 Contribution and Overview 16

2 Background   19
    2.1 Behavioral Modeling 19
    2.2 Operation Scheduling 21
       2.2.1 Basic Concept 21
       2.2.2 Previous Work on Scheduling 23
    2.3 Variable Lifetime 36
    2.4 Resource Allocation 38
       2.4.1 Basic Concept 38
       2.4.2 Previous Work on Allocation 41
    2.5 High-Level Synthesis Flow 48
    2.6 Testability Analysis 50
       2.6.1 Data Path Circuit Graph 50
       2.6.2 Sequential Path 50
       2.6.3 Sequential Loop 51

	2.7	Brief Review of High-Level Test Synthesis	52
		2.7.1 Built-In Self Test	52
		2.7.2 Scan	54

3 Sequential Depth Reduction During Allocation   57

3.1   Controllability and Observability Enhancement   58

3.2   Sequential Depth Reduction   60

3.3   Implementation   64

    3.3.1   Register Allocation   64

    3.3.2   Module Allocation   73

    3.3.3   Interconnection Allocation   74

3.4   Experimental Results   75

3.5   Summary   81

4 Sequential Loop Reduction During Allocation   83

4.1   Effect of Sequential Loops on Testability   84

    4.1.1   Sequential Loops in Acyclic SDFG   84

    4.1.2   Sequential Loops in Cyclic SDFG   88

4.2   Implementation   93

    4.2.1   Register Allocation   94

    4.2.2   An Example   96

4.3   Experimental Results   97

    4.3.1   In the Non-Scan Environment   97

    4.3.2   In the Partial Scan Environment   99

4.4   Summary   105

5 Testability Synthesis During Scheduling   107

5.1   Scheduling for Controllability/Observability Enhancement   108

	5.2	Scheduling for Sequential Depth/Loop Reduction	110
	5.3	Implementation	112
		5.3.1 Algorithm of MPS	113
	5.4	Experimental Results	116
	5.5	Summary	121

6 Conditional Resource Sharing for Testability   123

	6.1	Hierarchical Control-Data Flow Graph	124
	6.2	Effect of Conditional Resource Sharing	129
		6.2.1 Conditional Resource Sharing for *ex5*	130
		6.2.2 Conditional Resource Sharing for *ex6*	135
	6.3	A Conditional Resource Sharing Method for HTS	136
		6.3.1 Register Allocation	137
		6.3.2 Allocation for an HB or EB: *RallocB*	137
		6.3.3 Allocation for a Layer: *Traversal*	139
		6.3.4 *RallocC* for Complete Allocation	141
		6.3.5 *RallocC* in the Partial Scan Environment	141
	6.4	Experimental Results	142
		6.4.1 In the Non-Scan Environment	144
		6.4.2 In the Partial Scan Environment	144
	6.5	Summary	148

7 State-of-the-Art High-Level Test Synthesis   151

	7.1	Synthesis for Built-In Self Test	151
		7.1.1 SYNTEST: Case Western Reserve University	153
		7.1.2 RALLOC: Stanford University	158
		7.1.3 SYNCBIST: University of California, San Diego	164
	7.2	Synthesis for Scan Path and Test Point Insertion	168

	7.2.1 Genesis: Princeton University	169
	7.2.2 BETS: NEC C&C Research Laboratories	176
	7.2.3 TBINET: University of Wisconsin	181
7.3	Test Synthesis at RT Level	186
	7.3.1 ADEPT: University of Illinois	187
	7.3.2 The Chen-Karnik-Saab Method: University of Illinois	190
7.4	Other Work	192
7.5	Summary	197
Index		199
Bibliography		215

Preface

Computer-aided design tools have been widely used for designing integrated circuits to achieve high component density, short time-to-market, and low cost. These tools mainly feature physical design for placement and interconnection routing, and logic design for functional modules and control units. Recently, a design technique that allows architectural optimization from a circuit's behavioral specification, referred to as high-level synthesis, has been gaining acceptance to cope with the ever-increasing design complexity at a more abstract level.

Testing of an integrated circuit is a process in which the circuit is exercised so that a defect can be exposed. As the integrated circuit gets more and more complex and inaccessible, it becomes more difficult and expensive to rigorously test the design.

Since high-level synthesis enables the designer to handle complex designs, testing needs to be considered with other objectives from the beginning of synthesis. However, most existing high-level synthesis techniques optimize the circuit architecture only for area and performance. Because testability is not considered at this earliest design stage, the optimized architecture can have serious testing problems which are too difficult and expensive to fix and manufacture.

The research reported in this book seeks to promote a better design approach based on high-level test synthesis (HTS). For a given test strategy to be used in the design, HTS is able to explore the synthesis freedom provided at high level to derive an inherently testable architecture at low or even no overhead. The book presents several effective HTS schemes for highly testable digital circuits, assuming non-scan or partial scan test strategy. These schemes are implemented in the Princeton HI-level Test Synthesis (PHITS) system to be presented in this book.

A survey of other representative works in HTS is presented as well to give readers a comprehensive understanding of this interesting area. Various HTS techniques are discussed for both scan and built-in self test methodologies. Register-transfer level test synthesis is also covered.

There have been increasing research activities in high-level test synthesis. This book can serve as a reference book for researchers or engineers who are interested in this topic. A self-contained introduction of high-level synthesis algorithms and digital testing is provided for readers with either synthesis or testing background so that they can easily follow the discussion of the book.

Chapter 1 introduces high-level test synthesis. Chapter 2 provides the background of high-level test synthesis terminology, operation scheduling and resource allocation algorithms used in high-level synthesis, and a brief review of previous HTS works. Chapters 3 to 6 focus on the discussion of PHITS test synthesis algorithms, especially on sequential depth reduction, sequential loop breaking, scheduling, and conditional branch. Chapter 7 surveys other representative HTS works.

The book is based mostly on my Ph.D. research work at Princeton University. So I would like to thank my advisors, Professor Niraj K. Jha and Professor Wayne H. Wolf, for their inspiration and guidance. I would also like to thank the following people for interesting discussions on testing and high-level synthesis which contributed to the book: Dr. Rabindra K. Roy (NEC C&C Research Laboratory), Dr. Sujit Dey (NEC C&C Research Laboratory), Dr. John M. Acken (CrossCheck Technology Inc.), Mr. Dhanendra Jani (Intel Corp.), Dr. Sandeep Bhatia (Cross-Check Technology Inc.), Dr. Jaushin Lee (Silicon Graphics Inc.), Prof. Kwang-Ting

Cheng (University of California, Santa Barbara), Prof. Edward McCluskey (Stanford University), and Prof. Yu-Chin Hsu (University of California, Riverside).

Finally, I wish to thank my parents and my wife Ai-lin for their love and support through the years. It is to them and my son Andrew that I dedicate this book.

Chapter 1

Introduction

Testing of a very large scale integrated (VLSI) circuit is a process that applies a sequence of inputs to the circuit and analyzes the circuit's output sequence to ascertain whether it functions correctly. As the chip density grows to beyond millions of gates, VLSI circuit testing becomes a formidable task. Vast amounts of time and money have been invested by the semiconductor industry just to ensure high testability of the products. A number of semiconductor companies estimate that about 7% to 10% of the total cost is spent in single device testing [115]. This figure can rise to as high as 20% to 30% if the cost of in-circuit testing and board-level testing is added. However, the most important cost can be the loss in time-to-market due to hard-to-detect faults. Recent studies show that a six-month delay in time-to-market can cut profits by 34% [115]. Thus, testing can pose serious problems in VLSI design.

Part of the reason testing costs so much is the traditional separation of design and testing. Testing is often viewed inaccurately as a process that should start only after the design is complete. Due to this separation, the designer usually has little appreciation of testing requirements, whereas the test engineer has little input into the design process. In order to effectively reduce testing cost, testing must be considered from the earliest design stages. However, there has been very little research effort to explore this idea.

On the other hand, as design complexity drastically increases, current gate-level synthesis methodology alone can no longer satisfy stringent time-to-market requirement. High-level synthesis [28, 45, 87, 88], which takes a behavioral specification as input and automatically generates a register-transfer level architecture, is hence considered a promising technology to boost design quality and shorten the development cycle. Existing successful production-use high-level synthesis systems include tools from system companies such as HIS [16] of IBM and CALLAS [22] of Siemens, and commercial products such as DSP Station from Mentor Graphics based on Cathedral [83, 129] and Behavioral Compiler from Synopsys [66, 82]. However, these systems focus mainly on area and performance optimization. Thus, the derived architecture can be very hard to test.

This book attempts to bridge the gap between testing and high-level synthesis to address the testing issue in the earliest design stage. The problem of synthesizing highly testable digital circuits from the high level, especially at the behavioral level, is examined. Several novel high-level test synthesis algorithms are proposed that can consider testability along with area and performance optimization. These algorithms exploit the freedom provided in data path synthesis during operation scheduling and resource allocation to derive an inherently testable architecture at little or even no area/delay overhead. These algorithms are implemented in a test synthesis system called PHITS. Experimental results of PHITS show that benchmark circuits can be synthesized for which high fault coverage in a small amount of test generation time can be achieved, using the fewest registers and functional modules. This shows significant improvements over the high-level synthesis algorithms which do not consider testability.

A survey of other current high-level test synthesis systems is also presented to provide the readers with a comprehensive understanding of this important research area.

1.1. CIRCUIT TESTING

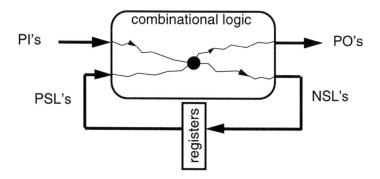

Figure 1.1: Generic model of a sequential circuit with the stuck-at fault denoted as a dot.

This introductory chapter is organized as follows. The basic concept of circuit testing and the role of high-level synthesis in VLSI design are explained first. Then the importance and advantage of considering testing during high-level synthesis is established. The chapter concludes with the contribution of the book and the organization of the remainder of the book.

1.1 Circuit Testing

VLSI circuit testing consists of applying a sequence of inputs to the circuit and comparing the output sequence with an expected output sequence. Any difference from the expected output sequence is said to have an *error*. The cause of an error is said to be a *fault*. Only the single stuck-at fault model is considered in this book for testability evaluation, which assumes at most one signal in a circuit is stuck at a logic value 0 or 1. We define *fault coverage* as the number of detected faults divided by the total number of faults, and *fault efficiency* as the number of detected faults and redundant faults divided by the total number of faults.

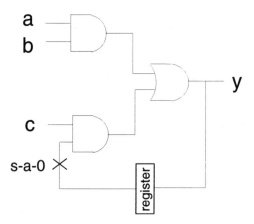

Figure 1.2: Example of a sequential circuit with a stuck-at-0 fault.

A generic model of a sequential circuit depicted in Figure 1.1 is used to illustrate the testing problem. The sequential circuit consists of combinational logic and registers. The registers are the memory elements to store the circuit state. The combinational logic takes the input values from the primary inputs (PIs) as well as the present state in the registers from the present state lines (PSLs), and generates the output values at the primary outputs (POs) as well as the next state on the next state lines (NSLs) for the registers.

In order to detect a fault in the combinational logic of a sequential circuit, one approach is to apply a deterministic automatic test pattern generation (ATPG) tool to sensitize a path from the primary inputs through the registers to activate the fault, and propagate the error effect through the registers to the primary outputs, as shown in Figure 1.1. Therefore, a sequence of vectors is required to test a fault. For example, in order to detect the stuck-at-0 fault on the present state line marked in the sequential circuit shown in Figure 1.2, first the primary inputs a and b must be set to both 1's so that the register can be set to 1. Then, if the stuck-at-0 fault

1.1. CIRCUIT TESTING

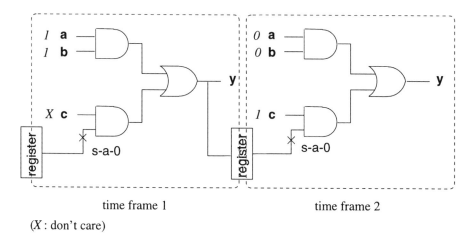

(X: don't care)

Figure 1.3: Iterative array model expanded into two time frames to detect the fault in the circuit of Figure 1.2.

exists, the present state line cannot be set to 1 by the register. This error effect 0 can be propagated to the primary output y by another test vector, where $a = b = 0$, and $c = 1$. Hence, totally a sequence of two vectors is required to detect the fault for this small example.

The above test generation procedure also indicates the underlying *iterative array model* [3] that most sequential ATPG tools are based on. In this model, the combinational logic block of the sequential circuit unter test is duplicated in each time frame. Figure 1.3 illustrates the iterative array model expanded into two time frames for testing the circuit in Figure 1.2. In time frame 1, test vector (a, b, c) $= (1, 1, X)$ is generated to activate the stuck-at-0 fault, where X denotes "don't care". In time frame 2, test vector $(0, 0, 1)$ is generated to propagate the error to the primary output y. In general, time frame expansion by sequential ATPG involves an intensive searching process within each time frame as well as across time frames in order to find the test sequence. For large sequential circuits, such search process by ATPG is computationally expensive.

The high sequential ATPG complexity arises due to poor controllability and observability of the registers, or memory elements. Controllability is the ability to set each line in the circuit to a specific value from the primary inputs. Observability is the ability to determine the value at any line in a circuit by controlling the primary inputs and observing the primary outputs. Important structural properties of sequential circuits that contribute to high ATPG complexity include [3, 33]:

- *sequential depth*: the number of register levels on a signal path between the primary inputs and the primary outputs.

  It can be shown that the length of a test sequence is linear in terms of the sequential depth [90].

- *sequential loop*: a cyclic signal path that involves registers.

  It can also be shown that the length of a test sequence is exponential in terms of the length of a sequential loop [62].

If the sequential ATPG uses the time frame expansion method based on the iterative array model, the number of processed time frames is equal to the test length. The longer the test length is, the longer the ATPG time is, and the more difficult it is for ATPG to find a test sequence under a practical processing time limit. Therefore, to alleviate the difficulty with sequential testing, one practical solution is to address the structural problem caused by sequential depths and loops by applying high-level test synthesis to derive the architecture with good controllability and observability. Note that recent work [85] that investigates another circuit attribute, termed *density of encoding*, demonstrates the effect of different state encodings on ATPG complexity. Since the major techniques are gate-level test synthesis, it is included in our discussion. Readers can refer to [85] for details.

Another circuit testing technique, called *built-in self test* (BIST) [12], is widely used where the circuit has the capability of testing itself with test vector generation

1.2. HIGH-LEVEL SYNTHESIS IN VLSI DESIGN

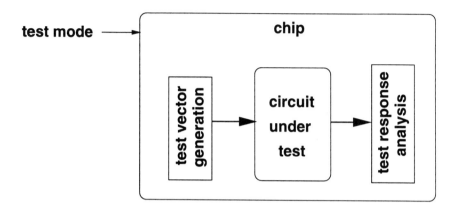

Figure 1.4: Generalized BIST configuration with test vector generation and test response analysis embedded in the chip.

and test response evaluation hardware embedded in the circuit, as shown in Figure 1.4. The major advantage of BIST over the ATPG-based approach is the cost elimination/reduction in deterministic ATPG and external test equipment used for test application. However, BIST hardware may impose larger area and delay overhead on the design, and cannot guarantee good fault coverage for general circuits. This book will focus first on developing high-level test synthesis techniques based on deterministic ATPG, as will be presented in Chapters 3 to 6. The major works on high-level synthesis for BIST testability will then be overviewed in Chapter 7. A complete discussion of high-level test synthesis for both test strategies is therefore provided.

1.2 High-Level Synthesis in VLSI Design

Conventional register-transfer level (RTL) synthesis or logic synthesis starts with a given RTL architecture where scheduling and allocation are already determined. So, if the result of RTL/logic synthesis cannot satisfy design requirement, one simple way to improve the design quality, usually by at most 10%, is to just modify the

RTL/logic synthesis script to guide the synthesis process under tighter constraints. However, if such 10% improvement is still not satisfactory, a more effective way to meet the design requirement is to directly modify the starting RTL architecture. This would require the designer to manually change the scheduling and allocation decisions for the RTL architecture and estimate its design quality, which is very time-consuming and will become the bottleneck for large and complex designs in the conventional design flow. Figure 1.5 shows such RTL/logic synthesis flow.

High-level synthesis can facilitate conventional synthesis methodology to remove the time-consuming bottleneck by performing scheduling and allocation to explore a circuit's design space, such that a set of RTL architectures for the circuit is derived automatically. Figure 1.6 illustrates this concept, in which a set of RTL architectures is produced and evaluated automatically by high-level synthesis. The designer can then select the most desirable design point to start with for RTL/logic synthesis. If the behavioral specification has to be modified in order to satisfy design requirement, it is a much easier task than modifying the RTL description because the behavioral description has a higher level of abstraction.

Research in VLSI computer-aided design has made high-level synthesis possible from an abstract behavioral specification to a complete hardware implementation. Figure 1.7 shows the overall flow of VLSI synthesis. The circuit's functionality is specified in a program-like description, called a behavioral specification. In the behavioral specification, the cycle-by-cycle behavior and hardware resource requirement are not determined yet. Since such an abstraction level is much closer to human reasoning, the behavioral specification is easier for the designers to write and analyze. Furthermore, starting with the behavioral specification, high-level synthesis can quickly trade off different objectives for complex designs during architecture optimization. Therefore, high-level synthesis has started gaining practical acceptance [16, 22, 77, 83, 106, 116, 120].

1.2. HIGH-LEVEL SYNTHESIS IN VLSI DESIGN

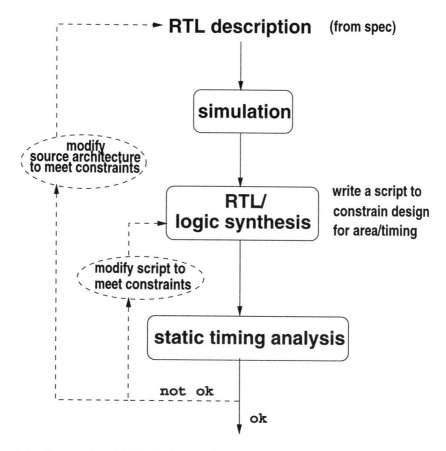

Figure 1.5: Conventional RTL/logic synthesis flow, where modifying the source architecture can be very time-consuming.

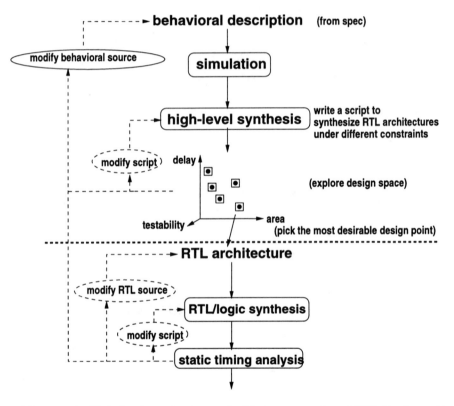

Figure 1.6: High-level synthesis can facilitate conventional RTL/logic synthesis.

1.2. HIGH-LEVEL SYNTHESIS IN VLSI DESIGN

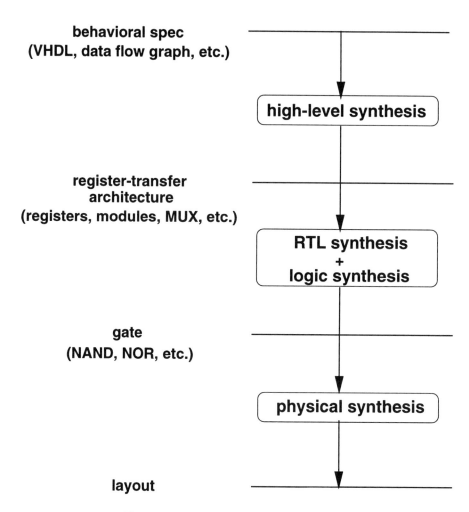

Figure 1.7: Overall VLSI synthesis flow.

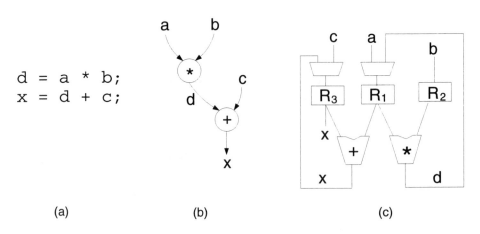

Figure 1.8: (a) Behavioral specification; (b) its data flow graph; (c) one possible register-transfer architecture.

Figure 1.8(a) shows an example of a circuit's behavioral specification, where each variable represents a signal name and each operation represents the computation it performs. This program-like specification is usually transformed into an equivalent graphical representation, called a **data flow graph** (DFG), where each node is associated with an operation and each arc is associated with a variable. Figure 1.8(b) shows the derived data flow graph for the behavior specified in Figure 1.8(a).

Given a behavioral specification, high-level synthesis [28, 45, 87, 88] then translates it into a register-transfer architecture which has fixed cycle-by-cycle behavior while satisfying cost constraints such as area, performance, and testability. Figure 1.8(c) depicts one possible register-transfer architecture that realizes the behavior specified in Figure 1.8(a). In this architecture, three registers are allocated by high-level synthesis to store the values of the variables: R_1 for a and d, R_2 for b, and R_3 for c and x. One multiplier and one adder are mapped to perform the specified operations. The execution time of each operation is also scheduled by high-level synthesis. In the first cycle, the multiplier takes a and b as operands from R_1 and

1.3. HIGH-LEVEL TEST SYNTHESIS

R_2, respectively, to generate the product d in R_1. In the second cycle, the adder takes c and d as operands from R_3 and R_1, respectively, to generate the sum x in R_3.

After the register-transfer architecture is derived, RTL synthesis and logic synthesis [24, 25, 88, 117] generate a netlist of logic gates implementing the architecture and further optimizes it. Physical synthesis [112, 118] then performs place-and-route to make the final layout. In this book, we will focus only on the earliest design stage in the synthesis flow—high-level synthesis—and demonstrate the advantage of considering testability at the earliest design stage.

Notice that RTL synthesis is used to produce an initial logic-level design from a register-transfer description before logic synthesis starts to optimize the boolean network. Since the register-transfer architecture already has the resource allocated and the operations scheduled, RTL synthesis lacks the full capability of performing architectural optimization for different objectives. However, some people still consider that RTL synthesis shares certain common synthesis techniques and objectives with high-level synthesis. This is because although conventional allocation and scheduling are done by high-level synthesis, some optimization/refinement on the derived architecture can still be performed by RTL synthesis to improve the architecture's quality. Therefore, this book includes the discussion of major works on RTL synthesis for testability improvement in Section 7.3.

1.3 High-Level Test Synthesis

Most of the existing high-level synthesis algorithms, to be reviewed in Chapter 2, are for area and delay optimization. Area is measured by counting registers, modules, and multiplexer inputs. Delay is measured by critical path delay in clock cycles. Since testability is not considered by these schemes, the optimized architecture can have very good tradeoff in area and performance, but poor testability, making it very hard to test and manufacture. Although RTL or lower-level design-for-test (DFT) techniques [4] can then be applied to improve testability, the improvement is restricted by the given RTL architecture. Therefore, in order to achieve high testability, a considerable amount of area and delay overhead can be incurred by the DFT methods. Furthermore, high-level synthesis can explore a larger design space than lower-level synthesis. An inherently testable architecture may already exist in the design space, which can be derived by high-level synthesis to produce a highly testable circuit at low or even no area/delay penalty.

The above concept is illustrated by the hypothetical three-dimensional design space shown in Figure 1.9, where each design point denotes an architecture optimized with different emphases on area, delay, or testability. The white design point represents an architecture optimized in testability as well as area and delay. So it can have better testability than the black design point, which disregards testability during synthesis. Although adding DFT circuitry to the black design point can enhance testability, as indicated by the gray design point, it may also result in larger area/delay in the final circuit than the white design point.

Consequently, high-level test synthesis (HTS), which can consider testability along with the other objectives at as early as the high-level design stage, is a critical component of a useful synthesis system. Assuming *a priori* no test strategy, or some test strategy such as scan or built-in self test, an HTS system aims to derive an inherently testable architecture at minimal area/delay overhead. Since such a testable architecture requires far less or even no test investment at lower levels, its final implementation can have better overall area and performance than implementations made testable by lower-level DFT methods.

1.3. HIGH-LEVEL TEST SYNTHESIS

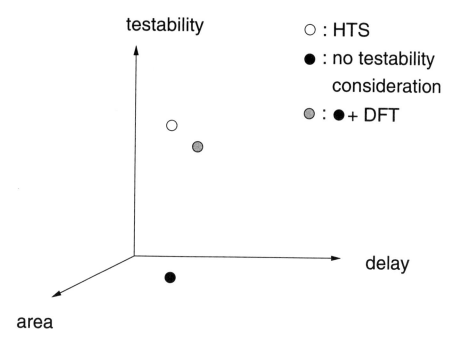

Figure 1.9: Hypothetical three-dimensional design space considering area, delay, and testability.

1.4 Contribution and Overview

This book first presents a novel high-level synthesis system, called the *Princeton HI-level Test Synthesis* (PHITS) system, for data path testability optimization, assuming a two-level testability strategy using non-scan and partial scan. PHITS can exploit the freedom provided during data path synthesis by operation scheduling and resource allocation to derive an inherently testable architecture at low or even no area/delay overhead.

Also reviewed are a number of other state-of-the-art test synthesis systems to provide readers with a comprehensive survey of high-level test synthesis.

The major contribution of this book, which will be illustrated in detail in the following chapters, can be briefly described as follows:

- A high-level synthesis approach for testability is presented, which can consider controllability and observability enhancement for registers. Assuming non-scan and partial scan test strategies, several heuristics based on this idea are proposed to derive inherently testable architectures at low or even no area/delay overhead.

- Testability is evaluated by deterministic ATPG to study the efficacy of the proposed approach, rather than by only counting the sequential loop structures.

- The experimental results show the advantage of our HTS schemes over the other lower-level DFT methods, in terms of the number of scan registers, the fault coverage, and the test pattern generation time.

- A comprehensive survey of representative high-level test synthesis works provides a self-contained reference to this topic.

1.4. CONTRIBUTION AND OVERVIEW

In Chapter 2, the basic definitions used in this book and a brief review of previous high-level synthesis works are provided. The allocation and scheduling algorithms exploited in the PHITS system are then explained in the next four chapters. Two allocation schemes in Chapters 3 and 4 are first introduced for test synthesis, which establish the importance of the scheduling method for testability presented in Chapter 5, although the PHITS system performs scheduling before allocation. A conditional resource sharing method for testability is then considered in Chapter 6.

Chapter 3 presents an allocation-for-testability scheme and its experimental results for simple data paths, assuming *a priori* no test strategy, or non-scan. Two synthesis rules are proposed based on enhancing controllability and observability of registers and reducing sequential depth in the circuit.

In Chapter 4, another allocation procedure based on sequential loop reduction is presented to take care of complex data paths that have loop constructs in the specifications. The procedure is differentiated into two levels of testability synthesis considerations with non-scan and partial scan assumptions, respectively.

Chapter 5 proposes a scheduling-for-testability algorithm, called *mobility path scheduling*, based on the allocation schemes developed in the previous two chapters. The mobility path scheduling algorithm can produce a better schedule where the allocation-for-testability schemes can then be easily applied.

Chapter 6 considers conditional resource sharing for testability during allocation. We present a more general behavioral model for conditional branches and a method for two-level testability synthesis.

Finally, Chapter 7 surveys other representative works on high-level test synthesis.

Chapter 2

Background

The key terms used in the book are defined in this chapter, specifically those in behavioral modeling, high-level synthesis, and testability analysis.

In high-level synthesis, scheduling and allocation are the two major tasks to perform. Scheduling determines the execution order of operations, while allocation assigns hardware to perform the operations. In general, the two tasks are NP-complete problems [46]. Therefore, there have been a number of heuristics developed to obtain optimal or nearly-optimal solutions by tackling problems one at a time. An overview of these heuristics is presented, since they will be applied extensively for test synthesis. Previous works on high-level test synthesis are briefly described as well.

2.1 Behavioral Modeling

The behavior of a circuit can be specified in a high-level hardware description language, and is usually translated into a graph form based on data flows and control flows defined in the specification. In a behavioral specification, a **basic block** is defined as a straight-line sequence of statements that contains no branches except at the very end [6]. Hence, a data flow graph (DFG) can be derived from the basic block where each node is associated with an operation and each arc is associated

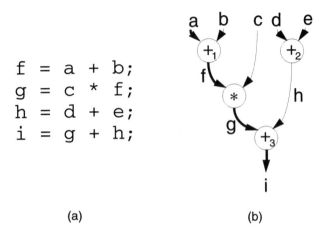

Figure 2.1: (a) Behavioral specification; (b) its data flow graph.

with a variable. Data dependency is implied by the direction of an arc. Figure 2.1(a) gives a simple example of a basic block with four statements, and Figure 2.1(b) shows its DFG.

The variables in a DFG can be classified into three subsets: primary inputs (V_I), primary outputs (V_O), and intermediate variables (V_M). A **data flow** in a DFG is denoted as $v_1 \xrightarrow{o_1} v_2 \xrightarrow{o_2} \cdots \xrightarrow{o_{n-1}} v_n$, where v_i, $1 \leq i \leq n$, is a variable in $V_I \cup V_O \cup V_M$, and o_j associated with each "$\rightarrow$", $1 \leq j < n$, is an operation (such as addition, multiplication, etc.) with v_j as one operand and v_{j+1} as the result. If v_1 and v_n are the same, the data flow is called a **cyclic data flow**, and v_1 or v_n is called a **boundary variable**. The set of all boundary variables in a DFG is denoted by V_B. A DFG is therefore said to be **acyclic** if it does not have any cyclic data flows; otherwise, the DFG is **cyclic**. For example, the thick arcs in Figure 2.1(b) indicate an acyclic data flow $a \xrightarrow{+_1} f \xrightarrow{*} g \xrightarrow{+_3} i$.

A cyclic data flow is usually imposed by a loop construct, such as *while* or *for*, in the behavioral specification. Figure 2.2(a) shows an example of a basic block

2.2. OPERATION SCHEDULING

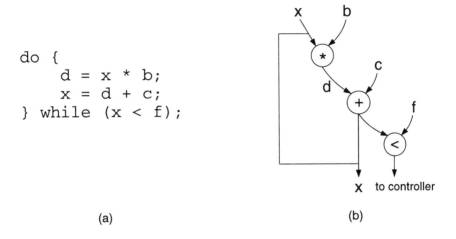

```
do {
    d = x * b;
    x = d + c;
} while (x < f);
```

(a)   (b)

Figure 2.2: (a) Basic block with a loop construct at the end; (b) its cyclic DFG with boundary variable x.

with a *while* statement at the end. Its DFG is depicted in Figure 2.2(b), which has a boundary variable x associated with the cyclic data flow $x \xrightarrow{*} d \xrightarrow{+} x$ imposed by *while*. Therefore, the current value of a variable on the cyclic data flow is dependent on its previous value in the last loop iteration. Notice that the data flow graph in Figure 2.2(b) does not describe the control flow, but only the data flows within *while*.

2.2 Operation Scheduling

2.2.1 Basic Concept

For each operation o associated with a node in a DFG, $o.earliest$ denotes the earliest possible cycle time o can execute in, and $o.latest$ denotes the latest possible cycle time o can execute in, without violating the data dependency defined in the DFG. Hence the **slack** of o is defined as the interval $[o.earliest, o.latest]$, and the **mobility**

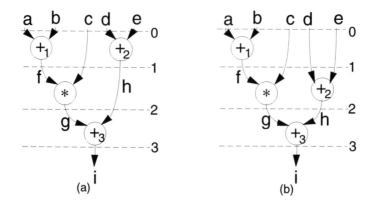

Figure 2.3: Two schedules for the same DFG: (a) S_a and (b) S_b.

of o is defined as $o.latest - o.earliest$. The sequence of operations with 0 mobility is called a **critical path**. For example, operation $+_2$ in Figure 2.1(b) has its slack [1, 2], so its mobility is 1. For operation $+_1$, $*$, and $+_3$, they have 0 mobility and are therefore on the critical path. Slack or mobility can be used to indicate the scheduling freedom for an operation.

Scheduling assigns an execution time, or control step [121], to each operation in a behavioral specification. In this book, we will only consider the **simple schedule**, where no operation is executed over multiple cycles [119]. Figure 2.3 shows an example of two different schedules, S_a and S_b, for the same DFG given in Figure 2.1(b), where the dashed line denotes the control step. The DFG is called a **scheduled data flow graph** (SDFG) after the schedule is determined. In this example, operation $+_2$ can be scheduled at either cycle 1 (by S_a) or cycle 2 (by S_b) without increasing the total execution time. However, S_a requires at least two adders because both $+_1$ and $+_2$ are executed at the same cycle 1 and cannot share a single adder, while S_b requires only one adder because only one addition is executed at each cycle. Figure 2.4 shows the scheduling effect on the final implementation.

2.2. OPERATION SCHEDULING

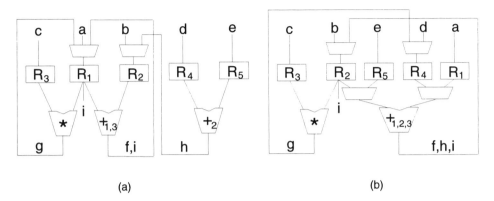

Figure 2.4: Two architectures derived respectively from the two schedules (a) S_a and (b) S_b.

The architecture in Figure 2.4(a) is derived from S_a with five registers, two adders, one multiplier, and two multiplexers. The architecture in Figure 2.4(b) is derived from S_b with five registers, one adder, one multiplier, and four multiplexers. They are two different architectures, but can perform the same behavior.

2.2.2 Previous Work on Scheduling

Most scheduling algorithms can be classified as transformational or iterative/constructive [87]. Other important algorithms include integer linear programming (ILP) scheduling, which is based on a mathematical formulation.

The transformational scheduling algorithm takes an initial schedule and attempts to perform behavior-preserving transformations to obtain an improved schedule. The *Yorktown Silicon Compiler* system [23] starts with a maximally parallel schedule which is then serialized to satisfy the resource constraint. The CAMAD design system [110] attempts to parallelize a fully serialized schedule by resource sharing.

The iterative/constructive approach schedules one operation or cycle at a time

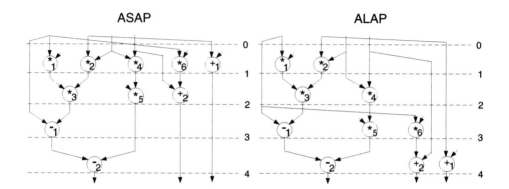

Figure 2.5: DFGs scheduled by ASAP and ALAP algorithms respectively.

until all the operations are scheduled. Important algorithms taking this approach are as-soon-as-possible (ASAP) scheduling, as-late-as-possible (ALAP) scheduling, list scheduling, critical path scheduling, force-directed scheduling, and path-based scheduling.

Popular scheduling algorithms based on iterative/constructive approach and ILP formulation are discussed in detail below.

2.2.2.1 ASAP and ALAP Schedulings

ASAP and ALAP schedulings are the simplest types of constructive approach. For each operation o in a DFG, ASAP algorithm schedules it at cycle $o.earliest$, while ALAP algorithm schedules it at cycle $o.latest$. ASAP scheduling is used by FACET [125], MIMOLA [86], and Flamel [122]. Figure 2.5 shows two DFGs scheduled by ASAP and ALAP algorithms, respectively. Although ASAP and ALAP algorithms can always assign the fastest schedule, they may require more hardware resource than the minimum. This is because the algorithms do not consider the critical path and cannot postpone the execution of the operation on the non-critical path to save resource. For example, for the ASAP scheduling given in Figure 2.5, four multipliers

2.2. OPERATION SCHEDULING

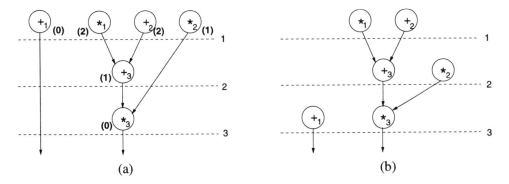

Figure 2.6: (a) DFG example where each operation has the number of operations between it and the end of the block as the priority (shown in parentheses); (b) list scheduling result.

are needed, because four multiplications are scheduled at cycle 1. However, if $*_4$ and $*_6$ on the non-critical paths are scheduled to cycle 2, and $*_5$ and $+_2$ are scheduled to cycle 3, only two multipliers are required without increasing the total number of execution cycles.

2.2.2.2 List Scheduling

List scheduling overcomes the problem with ASAP algorithm by maintaining a list of operations ordered by some priority function. Instead of selecting the operations as soon as possible, list scheduling schedules the next operation based on a priority. Therefore, the execution time of a lower-priority operation can be postponed when the resource limit is reached. Splicer [98] uses the mobility as the priority function. ELF [50] and Cathedral [83, 129] use the *urgency* of an operation as the priority function, which is the minimal number of clock cycles from that operation to the nearest constraint or output. The HAL system [108] can perform list scheduling with *force* as the priority, a concept that will be discussed in Section 2.2.2.4.

Figure 2.6(a) shows a DFG with the priority associated with each operation in the parentheses. The priority of an operation o used here is defined as the number

Table 2.1: List scheduling result in each cycle for the DFG in Figure 2.6(a) under the resource constraint of one adder and one multiplier.

cycle	ready list	priority list	list schedule
1	$(+_1, *_1, +_2, *_2)$	$(*_1, +_2, *_2, +_1)$	$*_1, +_2$
2	$(+_1, *_2, +_3)$	$(*_2, +_3, +_1)$	$*_2, +_3$
3	$(+_1, *_3)$	$(+_1, *_3)$	$*_3, +_1$

of operations between o and the end of the block in which o is located. We assume that the resource constraint is one adder and one multiplier. Table 2.1 lists for each cycle the ready list, the priority list, and the operations selected to schedule. The ready list maintained by list scheduling algorithm consists of the operations whose immediate predecessors are scheduled. These ready operations are candidates to be scheduled in the current cycle when priority and resource constraint are considered. For example, in the first cycle listed in Table 2.1, operations $+_1$, $*_1$, $+_2$, and $*_2$ are ready to be scheduled. But because of the resource constraint, only $*_1$ and $+_2$ with the highest priorities are selected to be scheduled in the first cycle. Similarly, in the second cycle, the ready list includes $+_1$, $*_2$, and $+_3$, where $*_2$ and $+_3$ with the highest priorities are selected to be scheduled. Finally, $*_3$ and $+_1$ are scheduled in the third cycle. Figure 2.6(b) shows the final list-scheduled DFG.

2.2.2.3 Critical Path Scheduling

Critical path scheduling used in MAHA [104] schedules the operations on the critical path first. For the remaining operations, the *freedom*, similar to the concept of mobility, is calculated. The freedom of an operation is calculated as the difference between the time when the input values are needed by the operation and the time when the result of this operation is required, less the propagation delay of the hardware. The operation with the least freedom is selected to be scheduled first.

2.2. OPERATION SCHEDULING

The MAHA scheduling algorithm, which is integrated with module allocation (to be discussed in Section 2.4), is outlined below:

- step 1: Given a DFG, set timing constraints and assign a delay value to each operation.

- step 2: Find the critical path and its total number of clock cycles required under timing constraints. If the timing constraints cannot be satisfied, go to step 1 with a new set of constraints.

- step 3: Allocate hardware modules to operations on the critical path.

- step 4: If there are no more operations to be scheduled, the algorithm terminates.

- step 5: Calculate/update the freedoms of all non-critical path operations.

- step 6: Schedule the operation with the smallest freedom.

    - If possible, a module that is already allocated is shared by this operation. Go to step 4.

    - Otherwise, either add another hardware module and go to step 4 or add one more cycle and go to step 2.

2.2.2.4 Force-Directed Scheduling

Force-directed scheduling in HAL [108] is a global load-balancing algorithm that attempts to distribute operations evenly over the cycles so that the resource utilization can be high. This scheme first derives a *distribution graph* for each type of operation to indicate the concurrency of the operations. It then calculates the *force*, a measure of concurrency distribution, on each operation and uses it to push the operation to a cycle where resource utilization is low.

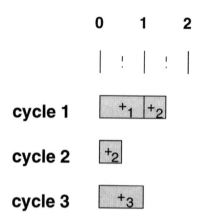

Figure 2.7: Distribution graph for the addition operations in the DFG in Figure 2.1.

A distribution graph is used to calculate the force for operations of the same type that could share a functional module. The distribution graph shows for each cycle how heavily loaded this cycle is with the operations of the same type, given that these operations are equally likely to be scheduled within their slacks. If an operation has its slack of k cycles, then $1/k$ is added to each cycle of the slack in the distribution graph. Figure 2.7 depicts the distribution graph for the addition operations of the DFG in Figure 2.1(b). Suppose the timing constraint on the total cycles is 3. Since $+_1$ and $+_3$ must be scheduled in cycles 1 and 3, respectively, 1 is contributed to cycles 1 and 3 in the distribution graph, respectively. For $+_2$, since it could be scheduled in either cycle 1 or cycle 2, $1/2$ is contributed to each cycle.

Given a distribution graph, for each possible schedule of an operation o to cycle i, a force is calculated as

$$F_i = \sum_{j \in [o.earliest, o.latest]} DG_j \times \Delta(j, i)$$

where j is in the range of os slack, DG_j is the value of the distribution graph in cycle j, and $\Delta(j, i)$ is the change in the value of the distribution graph if the schedule of o is fixed to cycle i. For example, in Figure 2.7, the force involved in assigning $+_2$

2.2. OPERATION SCHEDULING

to cycle 1 is $F_1 = 1.5 \times 0.5 + 0.5 \times (-0.5) = 0.5$. The first term accounts for the change involved in adding $+_2$ to cycle 1, since its probability increases from 0.5 to 1, while the second term accounts for removing $+_2$ from cycle 2, which decreases the probability from 0.5 to 0. The positive force F_1 indicates $+_2$ should not be scheduled to cycle 1, because cycle 1 is heavily loaded with addition. Similarly, we can calculate the force to schedule $+_2$ in cycle 2 as $F_2 = 1.5 \times (-0.5) + 0.5 \times 0.5 = -0.5$. Once all the forces are calculated, the schedule with the least force, or F_2 in this case, is selected. So $+_2$ is scheduled to cycle 2 and the distribution graph is updated to reflect this decision.

2.2.2.5 Path-Based Scheduling

Path-based scheduling [27] minimizes the number of clock cycles on all execution paths of a design description by exploiting the conditional branches. Instead of using a data flow graph, path-based scheduling algorithm operates on the *control flow graph* (CFG) of a design description to derive a schedule. In a CFG, the nodes represent operations to be scheduled, and the arcs give the precedence relation between nodes. That is, for an arc from nodes i to j, js operation can be executed after is operation is executed. If a node has more than one successor, this node is said to be a *conditional branch*.

Figure 2.8 shows an example of CFG [15] where node 2 is a conditional branch with signal **branch** as the predicate to be tested. Nodes 3, 4, 5, and 6 are taken if **branch** is true; otherwise, nodes 10 and 11 are taken. The feedback arc from nodes 9 to 1, which indicates iterative execution, is broken to make the CFG acyclic for path-based scheduling. Note that since the actual operations represented by the nodes are not of interest in our discussion, they are not explicitly shown in the figure.

Constraints are computed between two nodes that must be scheduled into two

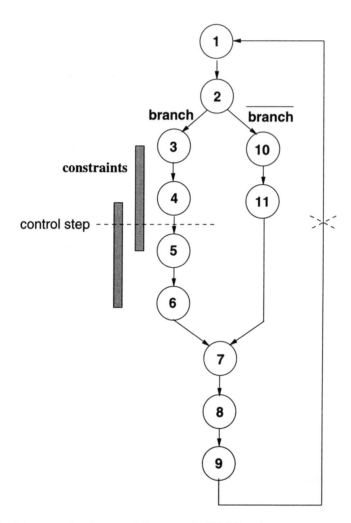

Figure 2.8: An example of control flow graph (CFG), where constraints are represented by intervals. Note the feedback arc is broken to make the CFG acyclic for path-based scheduling algorithm.

2.2. OPERATION SCHEDULING

different cycles. For example, in the CFG of Figure 2.8, if the operations of nodes 3 and 5 both assign values to the same signal, they cannot be scheduled into the same cycle. This constraint is represented by an interval from node 3 to node 5, as indicated by a bar in Figure 2.8. Other constraints can be derived as well, such as an IO port signal cannot have both read and write operations at the same time, and operations sharing the same module cannot be executed in the same cycle. If two constraints have their intervals overlapped, both constraints can be satisfied by introducing a control step in the overlapping sub-interval. For instance, in Figure 2.8, a control step is put between nodes 4 and 5, which satisfies both constraints indicated by the two intervals.

It is important to minimize the number of clock cycles introduced to satisfy all the constraints for efficient scheduling. Path-based scheduling algorithm first schedules all the execution paths independently by using a clique partitioning technique to introduce the minimal number of cycles for each path. In Section 2.4.2.2, an effective clique partitioning heuristic will be introduced. Then the schedules found independently for the paths are overlapped in a way to minimize the number of cycles for all paths. The algorithm of path-based scheduling is outlined below with the CFG example in Figure 2.8 as an illustration:

- step 1: Given a CFG, make it acyclic and derive all the execution paths.

  Figure 2.9(a) depicts the two execution paths, *path*1 and *path*2, in the CFG of Figure 2.8.

- step 2: Map all constraints onto intervals that are applicable to each execution path.

  In Figure 2.9(a), constraints $C1$ to $C6$ are associated with *path*1, while $C7$ and $C8$ are associated with *path*2.

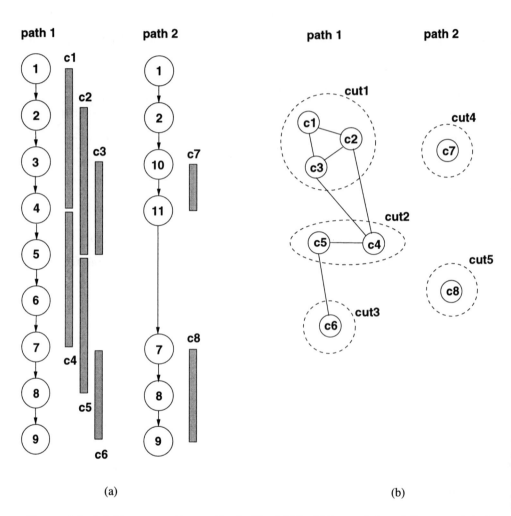

Figure 2.9: (a) Two execution paths in the CFG of Figure 2.8 and the associated constraints; (b) clique partitioning of the interval graph for each execution path.

2.2. OPERATION SCHEDULING

- step 3: Schedule each execution path independently, using the clique partitioning technique to introduce the minimal number of cycles.

  This is done as follows. In Figure 2.9(b), an *interval graph* is built for each path in Figure 2.9(a), where a node corresponds to an interval and an edge indicates that the two intervals corresponding to the two nodes connected by the edge are overlapped. A clique, which is a complete subgraph, contains all the intervals that overlap each other. So within the intersection of the intervals in a clique, referred to as the overlapping sub-interval, a control step can be placed to satisfy all the corresponding constraints. A "cut" is used to denote such an overlapping sub-interval associated with each clique. For example, the cut $cut2$ associated with the clique containing $C4$ and $C5$ is the sub-interval from nodes 5 to 7 in the CFG.

  As a result, a minimal clique partitioning of an interval graph gives the minimal number of cuts, which gives the minimal number of control steps, or clock cycles, for the corresponding execution path. Figure 2.9(b) shows the clique partitionings for $path1$ and $path2$. Figure 2.10(a) associates the cuts with each path.

- step 4: Overlap the schedules for all paths to minimize the total number of states using the clique partitioning technique.

  In Figure 2.10(b), a single interval graph is built for all the cuts in Figure 2.10(a). Again, in the interval graph, a node corresponds to a cut and an edge indicates the corresponding cuts are overlapped. Then a minimal clique partitioning of this interval graph gives the minimal set of cuts, denoted by fcs in Figure 2.10(b), that fulfills the schedule for all the paths, and thus the minimal number of total control states. Figure 2.10(c) shows the final schedule for the CFG in Figure 2.8 with the minimal number of control states.

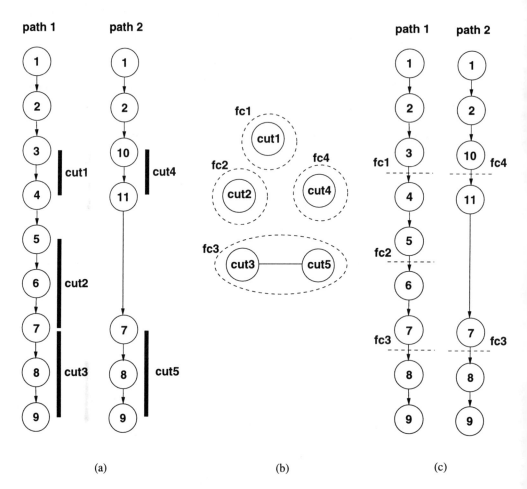

Figure 2.10: (a) Cuts found in Figure 2.9(b) associated with each execution path; (b) clique partitioning of the interval graph of cuts to derive the minimal number of control states; (c) final schedule with minimal control states.

2.2. OPERATION SCHEDULING

- step 5: Generate the control logic for the resulting schedule using conventional logic synthesis technique.

2.2.2.6 ILP Scheduling

ILP scheduling formulates the scheduling as an integer linear programming problem [76, 101]. A 0-1 variable x_{ij} is associated with each operation o_i and each possible cycle j in the slack of o_i. Each x_{ij} is set to 1 if o_i is scheduled at cycle j; otherwise it is set to 0. Since o_i can be scheduled in only one cycle in its slack, for each i, x_{ij}s are constrained so that exactly one x_{ij} can be 1. For example, if an operation o_1 can be scheduled at cycle 3, 4, or 5, the constraint $x_{13} + x_{14} + x_{15} = 1$ forces o_1 to be scheduled at only one of these cycles. Resource constraints and data dependency between two operations can be similarly expressed using the variable x_{ij}.

An example of of ILP scheduling formulation taken from [60] is used to illustrate the concept. Suppose the DFG has n operations to be scheduled into s cycles. A precedence relation between two operations o_i and o_j is denoted as $o_i \rightarrow o_j$, where o_i is the immediate predecessor of o_j. A total of m types of functional modules are available in the library. FU_k denotes the functional module of type k, while N_k denotes the number of functional modules available of type k. The relation $o_i \in FU_k$ denotes that FU_k can perform operation o_i. $[o_i.earliest, o_i.latest]$ is the slack of o_i.

The variables used in the formulation are:

- x_{ij}: as explained above.

- T_i: the cycle where o_i is scheduled.

- C_{step}: the total number of cycles required.

The problem can hence be expressed as

$$\text{minimize } C_{step} \tag{2.1}$$

subject to the constraints

$$\sum_{o_i \in FU_k} x_{ij} - N_k \leq 0, \text{ for } 1 \leq j \leq s, \ 1 \leq k \leq m \quad (2.2)$$

$$\sum_{j=o_i.earliest}^{o_i.latest} x_{ij} = 1, \text{ for } 1 \leq i \leq n \quad (2.3)$$

$$\sum_{j=o_i.earliest}^{o_i.latest} j \times x_{ij} - T_i = 0, \text{ for } 1 \leq i \leq n \quad (2.4)$$

$$T_i - T_j \leq -1, \ \forall o_i \to o_j \quad (2.5)$$

$$T_i - C_{step} \leq 0, \ \forall o_i \text{ without successors.} \quad (2.6)$$

The objective function 2.1 states that the goal is to minimize the total number of cycles. Constraint 2.2 means that in any cycle j, at most N_k operations of type k can be scheduled. Constraint 2.3 means o_i can be scheduled to any one cycle in its slack. Constraint 2.4 is to calculate the scheduled cycle T_i of o_i. Constraint 2.5 ensures the precedence relation. Constraint 2.6 states no operation can be scheduled after C_{step}.

The number of variables in the above ILP formulation grows in $O(n*s)$, while the number of constraints grows in $O(s*n+m)$. Because solving such a general ILP problem can be time-consuming, some heuristics, such as *zone scheduling* [60], have been proposed to reduce the problem size for efficient ILP scheduling.

2.3 Variable Lifetime

After scheduling, the DFG becomes a scheduled DFG, or SDFG. In an acyclic SDFG, the **birth time** of a variable v, denoted by $v.birth$, is the cycle time at which it is first defined; the **death time** of v, denoted by $v.death$, is the cycle time at which it is last used. The **lifetime** of v is therefore defined as the interval $[v.birth, v.death]$.

2.3. VARIABLE LIFETIME

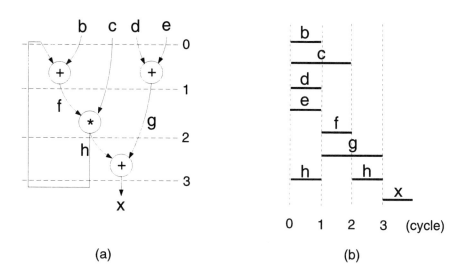

Figure 2.11: (a) Example of an SDFG; (b) its lifetime table.

In a cyclic SDFG, however, a variable may have multiple copies of identical lifetime for each loop iteration. To simplify the representation, all the feedback data flows at the loop boundary are broken to make the SDFG acyclic, and only the first loop iteration is considered to determine lifetime. For example, the lifetime of variable f in the cyclic SDFG in Figure 2.11(a) is defined as [1,2]. Furthermore, since the feedback data flow is broken, the lifetime of the associated boundary variable is split into two parts. For instance, the lifetime of boundary variable h in Figure 2.11(a) is defined as $[0,1] \cup [2,3]$.

A lifetime table of variables can be derived from an SDFG by performing lifetime analysis [6] on the SDFG. Figure 2.11(b) shows the lifetime table for the SDFG given in Figure 2.11(a). In a lifetime table, the cycle time is shown on the horizontal axis. The lifetime of each variable is denoted by a line segment whose left and right edges correspond to the birth and death times of the variable, respectively.

2.4 Resource Allocation

2.4.1 Basic Concept

Given a scheduled DFG, resource allocation assigns hardware elements like registers to store values, modules (or operators) to perform operations, and interconnection to carry signals, to produce a register-transfer architecture.

Register allocation, denoted by **R**, can be considered as a partition $\{R_1, R_2, \cdots, R_r\}$ of $V_I \cup V_O \cup V_M$ such that for any two variables v_i and v_j in R_k, $1 \leq k \leq r$, their lifetimes do not overlap, where V_I, V_O, and V_M are the sets of primary inputs, primary outputs, and intermediate variables, respectively. We can also define the birth time of R_i, denoted by $R_i.birth$, and its death time, denoted by $R_i.death$, as the youngest birth time and the oldest death time of the variables assigned to R_i, respectively. In addition, a register is called an input (output) register if a primary input (output) is assigned to it. If a register is both an input and output register, it is also called an IO register.

Similarly, module (or function unit) allocation, denoted by **M**, can be considered as a partition $\{M_1, M_2, \cdots, M_m\}$ of O, the set of all operations in a behavioral specification, such that for any two operations o_i and o_j in M_k, $1 \leq k \leq m$, their execution times do not conflict. We assume, for simplicity, that only the *regular module library* is used for module allocation. That is, each module in the library can perform exactly one operation.

From an SDFG, a undirected **module allocation graph** (MAG) is derived for module allocation, where each node corresponds to an operation in the SDFG, and each edge corresponds to the sharing compatibility for the same module between the two nodes it connects. The degree of sharing compatibility between operations for the same module can be further differentiated by assigning to each edge

2.4. RESOURCE ALLOCATION

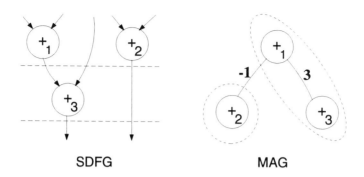

Figure 2.12: Example of a weighted MAG derived from an SDFG.

different values of *module sharing preference*, based on objectives such as area, performance, and testability. Therefore, a module allocation graph with the edges assigned module sharing preferences becomes a weighted module allocation graph. Module allocation can hence be obtained by finding sets of nodes in the module allocation graph where all of the members are connected to one another and the sum of the edge weights are the largest. This is the classical maximum-weight clique partitioning problem on the weighted module allocation graph [126].

An example of a weighted module allocation graph derived from an SDFG is given in Figure 2.12. Suppose two adders are allowed and, based on some area saving consideration, the module sharing preference is -1 between $+_1$ and $+_2$, and 3 between $+_1$ and $+_3$. Since the sharing preference between $+_1$ and $+_3$ is higher than that between $+_1$ and $+_2$, one adder will be shared by $+_1$ and $+_3$, and the other will be dedicated to $+_2$ only.

Wires are assigned by interconnection allocation to carry signals for variables among registers and modules. Similarly, two signals can be assigned to the same wire if the associated lifetimes do not overlap.

Figure 2.13 illustrates the allocation effect on the final implementation. Two

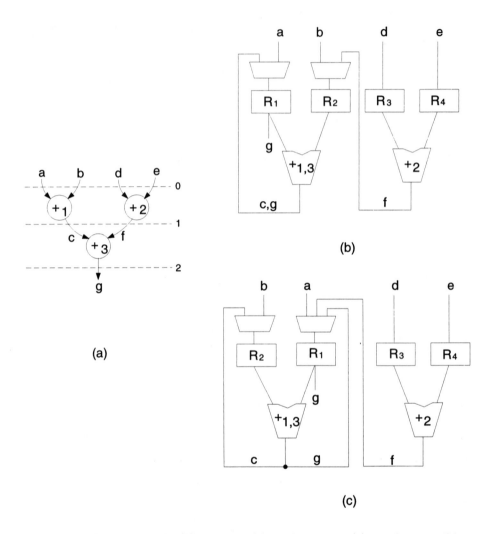

Figure 2.13: (a) An SDFG; (b) one possible architecture; (c) another possible architecture.

2.4. RESOURCE ALLOCATION

different architectures are allocated to realize the same behavior specified in the SDFG. Both architectures have four registers, but different variable assignments: in Figure 2.13(b), a, c, g are assigned to R_1, b, f, to R_2, d, to R_3, and e, to R_4, while in Figure 2.13(c), a, f, g are assigned to R_1, b, c, to R_2, d, to R_3, and e, to R_4. The architecture in Figure 2.13(c) requires one more interconnection and one more multiplexer input for signal g.

2.4.2 Previous Work on Allocation

Allocation algorithms can be classified as iterative/constructive or global [87]. The iterative/constructive approach allocates hardware elements one at a time to a variable, an operation, or an interconnection, while the global approach considers all elements for allocation at the same time. The EMUCS system [121] considers a global selection criterion, based on minimizing the number of registers, modules, and multiplexers needed, to allocate the next element. In the CHARM system [135], variables are iteratively assigned to share the same register if the cost of such register sharing is the lowest. Savings on modules and interconnections are also considered during evaluation of the sharing cost. Hence, module allocation and interconnection allocation can be performed at the same time as register allocation. The ELF system [50] selects all the operations in a cycle and allocates modules for them one at a time.

A number of popular allocation algorithms are discussed below in detail, including left edge algorithm, graph theoretic approach based on clique partitioning and bipartite matching, branch-and-bound method, and ILP formulation.

2.4.2.1 Left Edge Algorithm

The REAL program proposed by Kurdahi [71] is a constructive algorithm for register allocation. He showed that if a register is regarded as a wiring track and the lifetime of a variable as a wire segment, the greedy left edge algorithm (LEA) originally used for the channel routing problem [59] can be applied to allocate the minimum number of registers for an acyclic SDFG. Figure 2.14 illustrates an example of register allocation by the greedy left edge algorithm. In Figure 2.14(a), a lifetime table is first derived where all the variables are sorted by the birth time indicated by the left edge of each line segment. Then the greedy left edge algorithm starts register allocation at cycle 0 in Figure 2.14(b), scanning the lifetime table and placing the unassigned variables a and b in the first available time tracks corresponding to registers R_1 and R_2, respectively. At cycle 1, the algorithm assigns c and d to the first available time tracks in R_1 and R_2, respectively, and allocates two new registers R_3 and R_4 for e and f, respectively. At cycle 2, the algorithm assigns the remaining variables g and h again to the first available time tracks in R_1 and R_2, respectively. Thus, the greedy left edge algorithm allocates the minimum four registers for this example.

2.4.2.2 Graph Theoretic Approach

Two types of algorithms use graph theoretic approach for allocation: clique partitioning and bipartite matching.

Clique Partitioning FACET [126] performs allocation by applying heuristics to solve the clique partitioning problem on a *compatibility graph* for each register, module, and interconnection allocations respectively. In the compatibility graph for register allocation, a node corresponds to a variable, and an edge indicates that the variables corresponding to the two nodes connected by the edge do not have overlapping lifetimes. Figure 2.15 shows an example of allocating three registers to six variables using clique partitioning technique. Clique partitioning for module allocation and interconnection allocation can be formulated in a similar way.

2.4. RESOURCE ALLOCATION

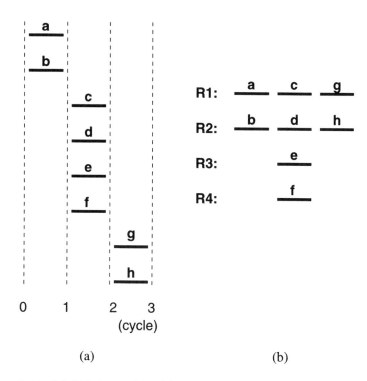

Figure 2.14: (a) Lifetime table; (b) register allocation by the greedy LEA.

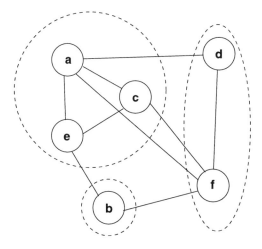

Figure 2.15: Clique partitioning for register allocation (variables in a clique share the same register).

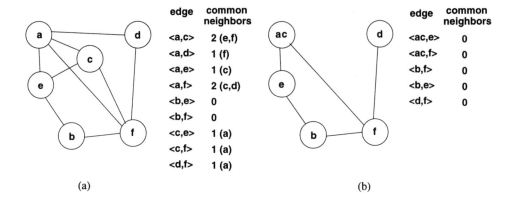

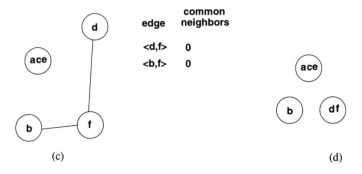

Figure 2.16: (a), (b), (c), and (d) are each iteration of the clique partitioning heuristic; the nodes in (d) correspond to the cliques in Figure 2.15.

2.4. RESOURCE ALLOCATION

Clique partitioning is a classic NP-complete problem. Tseng *et al.* [126] proposed the following heuristic to solve the problem for a given compatibility graph G:

- step 1: Select an edge $<v_i, v_j>$ in G such that v_i and v_j have the maximum number of common neighbors.

- step 2: Delete all the edges incident on v_i or v_j if they are not incident on the common neighbors of v_i and v_j.

- step 3: Merge v_i and v_j into a single node v_k.

- step 4: Go to step 1 until there is no edge left in G.

Each node in the final graph corresponds to a clique.

The algorithm can be illustrated in Figure 2.16 by clique-partitioning the graph in Figure 2.15. The common neighbors of the nodes connected by an edge in the graph of Figure 2.15 are enumerated in Figure 2.16(a). There are several edges with the maximum number, 2, of common neighbors, and such a tie is arbitrarily broken by selecting edge $<a, c>$. So nodes a and c are merged and unnecessary edges are deleted. The result is a new graph as shown in Figure 2.16(b), where edge $<ac, e>$ is then selected to merge ac and e. The resulting new graph is shown in Figure 2.16(c), where edge $<d, f>$ is selected next to merge d and f. Figure 2.16(d) gives the final result, with the nodes corresponding to the cliques in Figure 2.15.

Bipartite Matching The LYRA & ARYL method [61] applies the maximum-weight bipartite-matching algorithm to allocate registers and modules while interconnection cost is considered. In a bipartite graph, the set of nodes is partitioned into two disjoint subsets, and each edge connects two nodes in different subsets. Figure 2.17 depicts an example of a bipartite graph for register allocation. One set

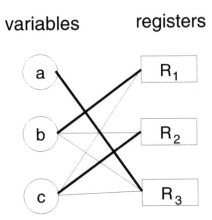

Figure 2.17: Example of a bipartite matching for register allocation.

of nodes represents variables with overlapping lifetimes, while the other set represents the registers. An edge is placed between a variable node and a register node if the variable can be assigned to the register. The allocation preference is assigned as the weight on the edge. Therefore, register allocation is transformed into the maximum-weight bipartite-matching problem, which requires to find a maximum-weighted subset of edges connecting each variable node to a unique register node. A polynomial time algorithm to solve this problem is presented in [102]. The thick edges in Figure 2.17 denote one solution to the maximum-weight bipartite-matching problem for register allocation in this example.

2.4.2.3 Branch-and-Bound Method

Splicer [97] uses a branch-and-bound search for resource allocation. The algorithm works cycle by cycle with a small number of look-ahead cycles to improve the branch-and-bound search quality in a reasonable amount of computation time. It can allocate registers and interconnections immediately after the operations scheduled at the same cycle have been assigned to the modules.

2.4. RESOURCE ALLOCATION

STAR [124] extensively uses sophisticated branch-and-bound searches in its two synthesis phases: data path construction and data path refinement. In the data path construction phase, a branch-and-bound search for the smallest increment of hardware cost associated with each binding decision is performed during register allocation and module allocation. In the data path refinement phase, the allocation quality is re-evaluated, and some of the binding decisions made during the data path construction phase is undone for refinement using a branch-and-bound search. The refinement process iterates until there is no more cost improvement.

2.4.2.4 ILP-Based Allocation

Hafer and Parker [54] used a linear programming technique for the whole data path synthesis, but it was only applicable to small examples. MIMOLA [86] and ADPS [101] used ILP methods for module selection at a lower computational complexity. The constraint used for module selection is formulated by setting the number of a certain type of module larger than or equal to the number of that type of operation scheduled in each cycle.

MIMOLA performs module allocation iteratively for the operations scheduled in the same cycle at a time. Its ILP formulation for such module allocation defines:

- n: total number of operations to be allocated.

- p: total number of modules available.

- x_{ij}: set to 1 if operation o_i is bound to module m_j; otherwise, set to 0.

- f_{ij}: defines feasibility of mapping o_i to m_j, which is set to 1 if the binding is feasible, or 0 if not.

- w_{ij}: cost of binding o_i to m_j, which is computed only if the binding is feasible.

The module allocation problem can be expressed as

$$\text{minimize} \sum_{i=1}^{n} \sum_{j=1}^{p} x_{ij} \times w_{ij} \qquad (2.7)$$

subject to the constraints

$$\sum_{i=1}^{n} x_{ij} \leq 1, \text{ for } 1 \leq j \leq p \qquad (2.8)$$

$$\sum_{j=1}^{m} f_{ij} \times x_{ij} = 1, \text{ for } 1 \leq i \leq n. \qquad (2.9)$$

The objective function 2.7 wants to minimize the total cost of module allocation. Constraint 2.8 states that no more than one operation in a cycle can be bound to the the same module. Constraint 2.9 says that each operation has to be bound to exactly one module.

2.5 High-Level Synthesis Flow

So far we have presented the major high-level synthesis techniques: operation scheduling, register allocation, module allocation, and interconnection allocation. The flow of such high-level synthesis steps in a particular design environment can vary from system to system based on different optimization and implementation considerations. The synthesis steps discussed above actually suggest a high-level synthesis flow where scheduling is performed before allocation, and allocation is performed for registers, then modules, and then interconnections. This order, depicted in Figure 2.18, is followed by our testability synthesis system PHITS, to be presented in Chapters 3 to 6.

To evaluate testability with PHITS, the output, a register-transfer architecture described in *BLIF* format, is further optimized by a sequential logic optimizer SIS [117]. A logic-level sequential ATPG tool STEED [48] (on SUN SPARCstation ELC) is then used to obtain testability.

2.5. HIGH-LEVEL SYNTHESIS FLOW

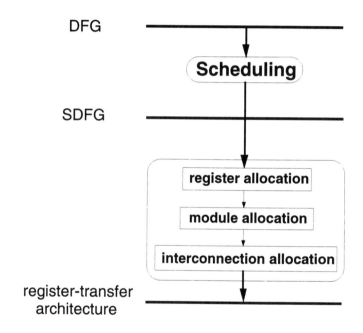

Figure 2.18: High-level synthesis flow in PHITS.

2.6 Testability Analysis

A data path circuit graph is proposed next to analyze the testability of an architecture during high-level synthesis.

2.6.1 Data Path Circuit Graph

After scheduling and allocation are done, a **data path circuit graph** (DPCG), similar to the logic-level *S-graph* proposed by Cheng and Agrawal [33], can be derived from the synthesized architecture, where each node represents a register and each arc represents a functional module connecting two registers. An arc from (to) a primary input (output) is also added to the node in the DPCG corresponding to the input (output) register. So the DPCG is an abstract representation of a circuit architecture. Figure 2.19(a) gives a circuit example with its DPCG shown in Figure 2.19(b).

2.6.2 Sequential Path

A **sequential path** between two nodes in a DPCG represents a connection between two registers in the circuit. The length of a sequential path is the number of arcs on the path in the DPCG. Because there may be more than one sequential path between a pair of input and output registers, we define the **sequential depth** between the register pair as the length of the shortest sequential path.

In the DPCG in Figure 2.19(b), the sequential path from R_1 to R_2 represents the register-transfer connection from R_1 through the adder to R_2 in the corresponding architecture in Figure 2.19(a). Its sequential depth is 1.

Since each node (arc) on a sequential path in a DPCG corresponds to a register (module) in the register-transfer architecture, a sequential path starting from a node for register R_1, passing through other nodes for register R_i, $1 < i < n$, and ending

2.6. TESTABILITY ANALYSIS

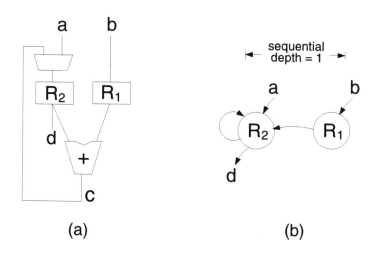

Figure 2.19: (a) Example of a register-transfer architecture; (b) its DPCG.

at a node for register R_n, can be denoted as $R_1^{V_1} \stackrel{M_1}{\to} R_2^{V_2} \stackrel{M_2}{\to} \cdots \stackrel{M_{n-1}}{\to} R_n^{V_n}$, where the superscripts V_j, $1 \leq j \leq n$, represents the set of variables assigned to R_j, while M_k, $1 \leq k < n$, associated with each "→" represents the module on the path receiving one operand from R_k and storing the result in R_{k+1}. These superscript V_js and M_ks can be omitted for simplicity if they are not of interest in some cases. Besides, the same term "sequential path" in a DPCG can also be used to refer to a physical register-transfer path in the architectural implementation without confusion.

2.6.3 Sequential Loop

A sequential path $R_1 \to R_2 \to \cdots \to R_n$ is called a **non-self-loop** if R_1 and R_n are the same, its length is larger than 1, and no register on the path is an IO register. If R_1 and R_n are the same but the length is 1, the loop is called a **self-loop**. In Figure 2.19(b), there exists a self-loop from R_2 back to R_2 in the DPCG. Because it is not difficult to control and observe values on a self-loop [33], we will focus only on the testability problems caused by non-self-loops, and use the term **sequential**

loop to mean a non-self-loop in the remainder of the book.

2.7 Brief Review of High-Level Test Synthesis

Little research attention has been paid to the area of high-level test synthesis. A brief description of the previous work on this topic is given next, based on two different test strategies, BIST and scan. A thorough survey of state-of-the-art high-level test synthesis systems will be presented in Chapter 7.

2.7.1 Built-In Self Test

Built-in self test, or BIST, is the capability of a circuit to test itself with little or no need for external test equipment or manual test procedures. On-chip circuitry is included to generate test vectors and to analyze output responses. For data path testing, BIST redesigns the registers in the circuit for test vector generation and test response analysis. The redesigned register is normally a built-in logic block observation (BILBO) register [67]. It supports BIST using several different modes of operation depicted in Figure 2.20: normal, serial scan in/out, random test pattern generation (RTPG), and multiple-input signature register (MISR). During BIST operation to test the combinational logic module, the register at the input to the module is set to RTPG mode to generate test patterns, while the register at the output of the module is set to MISR mode to compress the test response. Then, during serial scan in/out mode, the compressed test response, or signature, is scanned out for observation, and the seed for random test pattern generation is scanned in for the next session of BIST operation.

There is a restriction on such BIST strategy that is caused by some register whose input and output are directly connected by the same functional module. Such a register is called a *self-adjacent register*, and the feedback connection imposed by

2.7. BRIEF REVIEW OF HIGH-LEVEL TEST SYNTHESIS 53

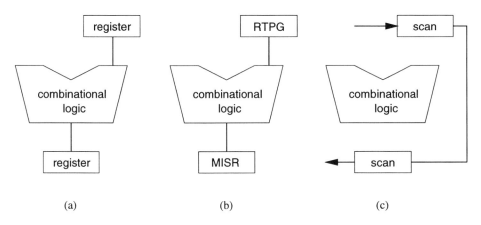

Figure 2.20: Different modes of operation in BIST: (a) normal operation; (b) BIST operation with random test pattern generation (RTPG) and multiple-input signature register (MISR); (c) scan in/out.

the self-adjacent register creates a self-loop. Figure 2.21 shows such an architecture with a self-adjacent register R_2. Hence, if BIST strategy is applied to test the ALU, R_2 has to perform as an RTPG and a MISR simultaneously, which is not supported by the BILBOs operation modes. In this case, a concurrent BILBO (CBILBO) register [133] which can perform both modes simultaneously is required to break the self-loop, which, however, will impose larger area overhead.

To satisfy the above restriction, Avra [9] proposed a graph-theoretic algorithm for register allocation to minimize the number of self-adjacent registers generated. For those cases where self-adjacent registers remain unavoidable, CBILBO registers are used. This work is discussed in greater detail in Section 7.1.2.

Papachristou *et al.* [100] developed an allocation scheme using ILP technique to derive an architecture with no self-loops. They also inspected random pattern testability of each module based on its functionality to further trade off testing time and hardware overhead. Moreover, Papachristou [99] proposed several rescheduling transformations to locally improve a given initial schedule such that the data path

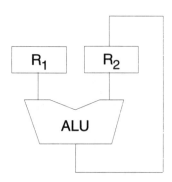

Figure 2.21: Architecture with a self-loop imposed by self-adjacent register R_2.

allocation algorithm in [100] is more applicable for BIST. His later test synthesis system SYNTEST [56] will be discussed in Section 7.1.1.

Gebotys and Elmasry [47] presented a simple method for BIST test strategy that uses a synthesis search based on area and delay constraints followed by a testability search based on area, delay, and test cost constraints. The method either finds a solution, or returns to the synthesis search for a more testable solution.

2.7.2 Scan

In the scan test strategy [134], the circuits registers are implemented by scan registers which at scan mode are interconnected into a shift structure so that it can scan in a test vector and scan out the test response. Since the scan register performs only the normal mode and the serial scan in/out mode, its area overhead is much smaller than that of a BILBO register. Scan strategy can be classified as full scan [134], which scans all the registers, or partial scan [5, 8, 29, 33, 35, 36, 72, 123], which scans only a subset of registers to further reduce overhead.

The testability synthesis method presented above by Gebotys and Elmasry [47] also considers multiple scan chains for full scan during allocation.

2.7. BRIEF REVIEW OF HIGH-LEVEL TEST SYNTHESIS

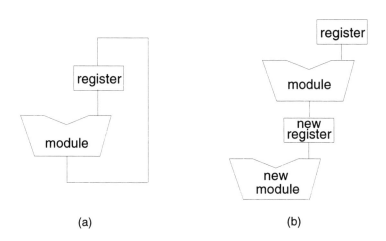

Figure 2.22: (a) Architecture with a self-loop; (b) the self-loop is broken by diverting a value to a new register.

Partial scan is first considered in the high-level test synthesis scheme proposed by Mujumdar et al. [92] where a network flow model is used for resource allocation to reduce the number of self-loops. A self-loop is created when a value is transferred from a module to a register in one cycle, and from the register to the same module in another cycle. Their allocation scheme checks to see if a self-loop exists between each register-module pair, and tries to eliminate it by either (i) diverting all the values going from a module to a register to other registers, or (ii) diverting all the values going from a register to a module to other modules. In case (i), the corresponding variables have to be re-assigned to other registers, whereas in case (ii), the corresponding operations have to be re-mapped to other modules. Figure 2.22 illustrates the idea for case (i) where the output of the module is diverted to a new register to break the self-loop. His later result on TBINET system [91] will be discussed in Section 7.2.3.

The PHITS system developed by Lee et al. [78, 79, 80, 81] considers both non-scan and partial scan test strategies, by reducing sequential depths and loops in a circuit during scheduling and allocation phases. The detail of PHITS test synthesis techniques will be presented in Chapters 3 to 6, while other test synthesis methods will be surveyed in Chapter 7.

Chapter 3

Sequential Depth Reduction During Allocation

Given a scheduled data flow graph (SDFG) of a circuit, the allocation process assigns hardware to synthesize a register-transfer architecture implementing the circuit behavior. However, if the allocation process does not consider testability, the derived architecture can have serious testing problems, which are very difficult and expensive to fix later at lower-level design stages.

In this chapter, we present an allocation scheme used by PHITS which can synthesize highly testable data paths without assuming *a priori* any test strategy. Two synthesis rules are proposed, one based on enhancing controllability and observability of registers, and the other based on reducing sequential depth between registers. These rules depend only on the register-transfer architecture of the implementation—they are independent of the test methodology used. Experimental results show that highly testable circuits can be synthesized by our algorithm for simple data paths. For complex data paths that contain cyclic data flows, one more testability synthesis rule is required which will be discussed in the next chapter.

3.1 Controllability and Observability Enhancement

The idea of enhancing controllability and observability of a circuit has been widely used in several logic-level DFT techniques [4]. The scan register structure is one such popular method employed in designing testable sequential circuits at the logic level. With enhanced controllability and observability of registers, this DFT technique makes the two major steps in test generation—fault excitation and error effect propagation—much simpler, and thus achieves high testability. However, scan registers can impose a considerable overhead in area and delay.

The above idea of enhancing controllability and observability can be actually considered at an even earlier design stage (i.e., the behavior level). Since the architecture of a circuit is determined by high-level synthesis, registers with good controllability and observability can be allocated during high-level synthesis to improve testability even without scanning any registers.

In high-level synthesis, register allocation assigns variables in the specification to registers in the register-transfer architecture. If a variable's value is not needed by all the computations at a certain cycle, we can reuse the register by storing another variable in it. Kurdahi and Parker [71] showed that the greedy left edge algorithm (LEA) can be used to allocate the minimum number of registers for an acyclic SDFG. The greedy left edge algorithm starts with the primary inputs born at cycle 0 for register allocation and proceeds forward to the last cycle, while a similar greedy right edge algorithm (REA), which treads the time axis in the reverse direction, starts with the primary outputs dead at the last cycle and proceeds backward to cycle 0. However, the greedy left edge algorithm does not consider testability and its effect on module and interconnection allocations that follow. The result may be an architecture with few registers but very poor testability.

3.1. CONTROLLABILITY AND OBSERVABILITY ENHANCEMENT

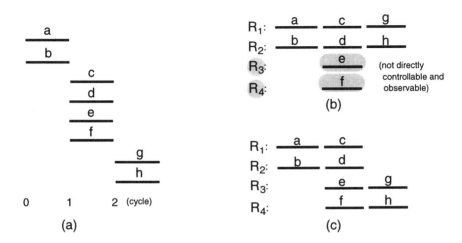

Figure 3.1: (a) Example of a lifetime table; (b) one possible register allocation by the greedy LEA; (c) another possible register allocation with testability consideration.

Figure 3.1(a) depicts an example of a lifetime table with eight variables, where the primary input set $V_I = \{a, b\}$, the intermediate variable set $V_M = \{c, d, e, f\}$, and the primary output set $V_O = \{g, h\}$. Its two possible register allocations are shown in Figures 3.1(b) and 3.1(c), where cycle times are on the horizontal axis and registers are on the vertical axis. Each line segment corresponds to a variable's lifetime—the variable is stored in a register for the time during which the variable's value is used.

If any one of the variables assigned to a register is a primary input (output) of the chip, this register is directly controlled (observed); if not, the register can be accessed only through other registers. Our goal is to ensure that as many registers as possible in the implementation are assigned at least one primary input/output variable. Therefore, it each row for a register is covered by a primary input (output), the corresponding register is directly controllable (observable). Figure 3.1(b) shows a minimum allocation of four registers by the greedy left edge algorithm for the variables in Figure 3.1(a): $\mathbf{R} = \{(a,c,g),(b,d,h),(e),(f)\}$. In this allocation, R_1 and R_2 are both controllable and observable, while R_3 and R_4 are not directly controllable and observable. Therefore, faults tested through R_3 and R_4 may be hard to test.

Suppose we know R_1 and R_2 are already easily observable through some other sequential paths even without primary output assignment of g and h. In this case, g and h can actually be reassigned to R_3 and R_4, respectively, to enhance the observability of R_3 and R_4 without increasing the number of registers. So the register allocation becomes $\mathbf{R}' = \{(a,c),(b,d),(e,g),(f,h)\}$, as shown in Figure 3.1(c). Now all the registers are either directly controllable or observable, which makes the circuit more testable. This example leads us to our first synthesis rule:

- **SR1**: Whenever possible, allocate a register to at least one primary input or primary output.

3.2 Sequential Depth Reduction

Since our first rule provides no testability suggestion for module allocation and interconnection allocation, it is not sufficient in itself to ensure high testability. A more general requirement of reducing sequential depth is therefore added to help ensure testability. This requirement is intimately related to register allocation and module allocation. For module allocation, several operations, such as additions and subtractions, may be implemented in a single module, such as an adder or ALU. The mapping of operations to modules, together with register allocation, determines the interconnection structure in the architecture, and therefore the sequential depth in the circuit as well.

Consider the SDFG of a circuit named *ex1* and the lifetime table for its variables, as shown in Figure 3.2, where $V_I = \{a,b,d,e\}$, $V_M = \{c,f\}$, and $V_O = \{g\}$. Suppose

3.2. SEQUENTIAL DEPTH REDUCTION

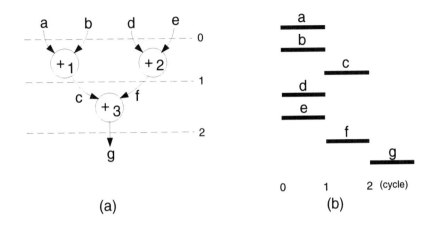

Figure 3.2: (a) SDFG of *ex1*; (b) its lifetime table.

by using SR1 we obtain a register allocation $\mathbf{R} = \{(a,c,g),(b,f),(d),(e)\}$, where each register is either controllable or observable. Also suppose two adders are allowed for the three additions $+_1$, $+_2$, and $+_3$.

There are two feasible module allocations: $\mathbf{M}_1 = \{(+_{1,3}),(+_2)\}$ or $\mathbf{M}_2 = \{(+_{2,3}), (+_1)\}$. Since in $\mathbf{M}_1$, operations $+_1$ and $+_3$ receive one of their operands (a and c, respectively) from the common register R_1, and also store their results (c and g, respectively) in the common R_1, $\mathbf{M}_1$ is preferred to $\mathbf{M}_2$ to save interconnection cost. After interconnection allocation is performed, which is straightforward in this example by using multiplexers, the final data path is implemented as in Figure 3.3(a). In this architecture, we can see that registers R_1, R_2, and the adder $+_{1,3}$ are not difficult to test because of relatively easy controllability at the input ports a, b and easy observability at the output port g. However, for registers R_3, R_4, and the adder $+_2$, although d and e are easy to control, the output of $+_2$ is hard to observe. An error effect needs to propagate through an additional register R_2 before it can be observed at g. The DPCG in Figure 3.3(b) reflects the analysis: the sequential

62 CHAPTER 3. SEQUENTIAL DEPTH REDUCTION DURING ALLOCATION

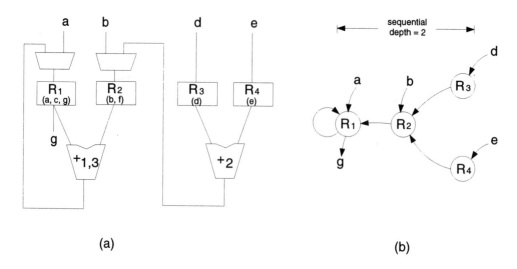

Figure 3.3: (a) One implementation of *ex1* without testability consideration; (b) its DPCG.

depths to the output register R_1 from R_1, R_2, R_3, and R_4 are 0, 1, 2, and 2, respectively. So the question arises: can we decrease the sequential depths from R_3 and R_4 to improve testability while keeping low area overhead?

Fortunately, we can reduce sequential depth with an area penalty that is usually small. Suppose the same module allocation $\mathbf{M}_1$ is used. Since the operations $+_1$ and $+_3$ are mapped to the same adder, the result c of $+_1$ can be effectively observable through the result g of $+_3$ at the same adder $+_{1,3}$. Thus it is not necessary to assign c to the output register R_1. This makes it possible to allocate R_1 to f, the result of $+_2$, together with g. Hence the circuit of $+_2$, and registers R_3 and R_4 that provide the two operands for $+_2$, can now be more easily observable.

Based on the above discussion, a new register allocation $\mathbf{R}' = \{(a, f, g), (b, c), (d), (e)\}$ is obtained by switching the assignment of c and f in $\mathbf{R}$ without increasing the number of registers. Notice that with $\mathbf{R}'$, the adder allocated to $+_1$ and $+_3$

3.2. SEQUENTIAL DEPTH REDUCTION

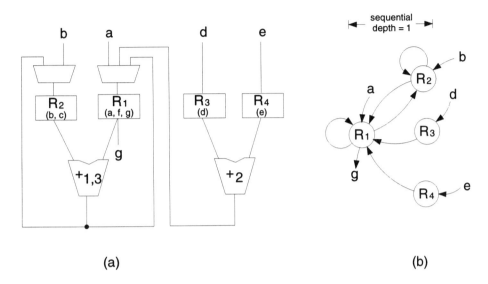

Figure 3.4: (a) Another implementation of *ex1* with testability consideration; (b) its DPCG.

can still access one operand (a or f) from the common register R_1, and the other operand (b or c) from the common register R_2, to save on interconnection.

The architecture is shown in Figure 3.4(a), where we can see that the value of c can be observed through the right branch from the output of adder $+_{1,3}$ to R_1, while the area overhead is only one more connection and one more multiplexer input. Figure 3.4(b) shows its DPCG where the sequential depths from both R_3 and R_4 to R_1 decrease from 2 to 1.

To verify the above idea, we applied a logic-level ATPG tool, STEED, to both circuits (implemented in 4-bit width). Our results show consistent improvement by proper allocation: the fault coverage for **R** is only 89.28%, while for **R'** it is 100%. The test generation time is 24.5 seconds for **R** and 10.7 seconds for **R'**. Besides, as we predicted, for **R** most undetected faults are aborted at the logic of R_3, R_4, and $+_2$. We can therefore describe the second synthesis rule SR2 as follows:

- **SR2**: Reduce the sequential depth from an input register to an output register.

3.3 Implementation

An effective allocation algorithm based on SR1 and SR2 is developed in the *Princeton HI-level Testability Synthesis* (PHITS) system [81]. PHITS can take an acyclic SDFG as the input, assuming all primary inputs are born at the same cycle and all primary outputs die at the same cycle. We also assume that the load signals of all registers and the select signals of all multiplexers are controllable due to our focus on testability of the data path only. As discussed before, because register allocation can drastically affect controllability and observability, PHITS first allocates registers while taking into account the effect of register choices on subsequent module and interconnection allocations. With the module sharing preference derived during register allocation, a weighted module allocation graph is built for module allocation. Interconnection allocation is performed last.

3.3.1 Register Allocation

Synthesis rules SR1 and SR2 are mainly embedded in the register allocation phase of PHITS. PHITS essentially takes an iterative/constructive approach to find register allocation for one variable at a time. A branch-and-bound version of the left edge algorithm is proposed as the core of register allocation. Register sharing preference among variables is developed at the same time during register allocation to guide the branch-and-bound search. Therefore, the branch-and-bound left edge algorithm can explore a larger search space and select a better binding variable than can the greedy left edge algorithm. For simplicity, in the algorithm described next, $R \leftarrow v$ denotes the assignment of variable v to register R, and $R_i \Leftarrow R_j$ denotes the merging

3.3. IMPLEMENTATION

of register R_j with another register R_i.

3.3.1.1 Branch-and-Bound Left Edge Algorithm

The branch-and-bound left edge algorithm (BBLEA) is shown in Figure 3.5. It takes as arguments two disjoint sets of variables: V_1, the variables with the same earliest birth time, and V_2, the variables born later. Since all the variables in V_1 have overlapped lifetimes, a new register must be allocated to each of them (lines 1-3). Those registers are only partially allocated at this moment. Then for each variable v in V_2, a partially allocated register, denoted as $R_{partial}$, with the closest death time to $v.birth$ is found (lines 5-6). The main difference between BBLEA and the greedy left edge algorithm is that for $R_{partial}$, BBLEA, instead of only searching for the first closest one, searches for all the unallocated variables with the same closest birth time to $R_{partial}.death$, and thus forms a search space V_S (line 8). It then selects the variable with the highest register sharing preference to assign to $R_{partial}$ (lines 9-10). The sharing preference function and the $branch\_and\_bound$ procedure will be discussed later in this section. Finally, after all the variables in V_2 are processed, the allocated register set **R** and the remaining unassigned variables in V_2 are returned (line 13). Similarly, a branch-and-bound right edge algorithm (BBREA) can be obtained by applying BBLEA to a flipped version of a normal lifetime table; that is, a version obtained by switching the birth time and the death time of each variable.

Given the lifetime table of an SDFG, the algorithm $Ralloc_{min}$ in Figure 3.6 can use BBLEA to allocate registers for all the variables in V_{in} cycle by cycle. The algorithm keeps partitioning the set V_{remain} of the remaining variables into two disjoint sets (lines 3-4) (i.e., V_1 and V_2) and applying BBLEA to them to allocate new registers (line 5), until V_{remain} becomes empty (line 2). Since the branch-and-bound search space defined in BBLEA only consists of those variables with the same

```
BBLEA(V₁, V₂) {
1.   foreach (vᵢ ∈ V₁) {
2.       Rᵢ ← vᵢ;
3.       R = R ∪ {Rᵢ};
4.   }
5.   foreach (v ∈ V₂) {
6.       R_partial = closest_register_to(v);
7.       if (R_partial == ∅) continue;
8.       V_S = find_search_space(R_partial);
9.       v_alloc = branch_and_bound(V_S);
10.      R_partial ← v_alloc;
11.      V₂ = V₂ − {v_alloc};
12.  }
13.  return(R, V₂);
}
```

Figure 3.5: Branch-and-bound left edge algorithm.

```
Ralloc_min(V_in) {
1.   V_remain = V_in;
2.   while (V_remain ≠ ∅) {
3.       V₁ = earliest_born(V_remain);
4.       V₂ = V_remain − V₁;
5.       (R_new, V_remain) = BBLEA(V₁, V₂);
6.       R = R ∪ R_new;
7.   }
8.   return(R);
}
```

Figure 3.6: Register allocation algorithm $Ralloc_{min}$.

3.3. IMPLEMENTATION

```
Ralloc() {
1.   (R, V_remain) = BBREA(V_O, V_M);   /* enhance observability */
2.   (R_temp, V_remain) = BBLEA(V_I, V_remain);  /* enhance controllability */
3.   R = merge(R, R_temp);
4.   if (V_remain ≠ ∅)
5.       R = R ∪ Ralloc_min(V_remain);
6.   return(R);
}
```

Figure 3.7: Register allocation algorithm *Ralloc* used in PHITS.

minimum gap of lifetime to a register, $Ralloc_{min}$ can allocate the same minimum number of registers as obtained by the greedy left edge algorithm.

3.3.1.2 *Ralloc*: Register Allocation in PHITS

Although algorithm $Ralloc_{min}$ can give the minimum number of registers, it does not consider testability sufficiently. To incorporate SR1 and SR2 in order to achieve testability, a modified version, *Ralloc*, is presented in Figure 3.7. This algorithm cannot guarantee minimality of the number of registers, but our experiments show that in almost all the cases it indeed allocates the minimum number of registers.

PHITS uses *Ralloc* to enhance observability first by applying BBREA to V_O and V_M, obtaining a register set $\mathbf{R}$ (line 1). Recall that V_O is on the right-hand side of a lifetime table, so BBREA is applied. This step allocates the output registers to as many intermediate variables as possible. PHITS then enhances controllability by applying BBLEA to V_I and the remaining unallocated intermediate variables, obtaining a temporary register set $\mathbf{R}_{temp}$ (line 2). This step allocates the input registers to as many of the remaining unallocated intermediate variables as possible. These two synthesis steps together enforce rule SR1. Then registers in $\mathbf{R}$ and $\mathbf{R}_{temp}$ are merged together if possible (line 3). Finally, if there are still unallocated

```
merge(R, R_temp) {
1.   foreach (tempR_i ∈ R_temp) {
2.       v = variable defining tempR_i.death;
3.       R_partial = closest_register_to(v);
4.       if (R_partial == ∅) continue;
5.       R_S = find_search_space(R_partial);
6.       tempR_merge = branch_and_bound(R_S);
7.       R_partial ⇐ tempR_merge;
8.       R_temp = R_temp − {tempR_merge};
9.   }
10.  foreach (tempR_i ∈ R_temp) {
11.      R_i = tempR_i;
12.      R = R ∪{R_i};
13.  }
14.  return(R);
}
```

Figure 3.8: Algorithm *merge* to merge $\mathbf{R}_{temp}$ with $\mathbf{R}$.

intermediate variables, $Ralloc_{min}$ is invoked to complete allocation (lines 4-5).

The *merge* algorithm is given in Figure 3.8. It is very similar to BBREA—it intends to allocate at a time a group of variables, which were previously assigned to a temporary register in $\mathbf{R}_{temp}$, to a partially allocated register in $\mathbf{R}$. For each temporary register $tempR_i$, the variable v defining its death time is first obtained (lines 1-2). Then a partially allocated register, denoted by $R_{partial}$, in $\mathbf{R}$ is found if the birth time of $R_{partial}$ is the closest to $v.death$ (line 3). Thereafter, $\mathbf{R}_S$, the search space for $R_{partial}$, is formed, and branch-and-bound search is performed in it (lines 5-6). The temporary register with the highest sharing preference is then merged with $R_{partial}$ (line 7). Notice that for simplicity, rather than all the variables in $tempR_i$, only the variable defining the death time for each temporary register in $\mathbf{R}_S$ is inspected for its preference for merging with $R_{partial}$. Finally, a new register is needed for each of the remaining non-mergeable temporary registers, because their

3.3. IMPLEMENTATION

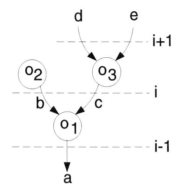

Figure 3.9: Example of sharing preference.

lifetimes are all overlapped by primary inputs (lines 10-12).

3.3.1.3 Sharing Preference for Variables and Operations

Sharing preference is used to guide branch-and-bound search in *branch_and_bound* procedure invoked in BBLEA and BBREA. Two types of sharing preferences are assigned during the search process: register sharing preference and module sharing preference. Since BBLEA and BBREA used in algorithm *Ralloc* assign variables to registers cycle by cycle, the register sharing preference to variables at the next cycle can be derived during register allocation at the current cycle. Therefore, when *branch_and_bound* is called for the variable in the next cycle, this preference can be applied in the same manner as in the previous cycle to guide register allocation.

Register sharing and module sharing are tightly correlated, so their preferences are derived simultaneously and used for register allocation. Later in module allocation, the same module sharing preference is also used as the edge weight in a weighted module allocation graph.

The sharing preference assignment in BBREA is illustrated in Figure 3.9. Suppose a, b, c, d, and e are variables, o_1, o_2, and o_3 are operations, and a is a primary output. Also suppose from the previous call of *branch_and_bound* that a is assigned to a register R and has a higher register sharing preference to b than to c. Therefore, with the search space of R formed by b and c, the next call to *branch_and_bound* will assign b to R, which makes the output of o_2 observable. To make the output of operation o_3 also observable, the technique used in the second implementation of circuit *ex1* can be applied. That is, o_1 and o_3 are mapped to the same module, if applicable, by assigning a relatively high module sharing preference for them. So c can effectively be observed through a, and the sequential depth through o_3 is reduced. The synthesis rule SR2 is thus enforced.

Interconnection complexity can be considered at the same time in a similar way as in Tseng and Siewiorek [126]. Since in the above example sharing o_1 and o_3 is preferred, it is desirable that their operands are also assigned to the common registers. Therefore, the register sharing preference for b and d, as well as c and e, can be assigned. Again, this preference can be used in a later call of *branch_and_bound* when d or e is targeted for register allocation.

The idea of deriving register sharing and module sharing preferences in BBLEA is similar. As for the *merge* procedure invoked in *Ralloc*, the sharing preference is already assigned during BBREA and BBLEA at lines 1 and 2 in *Ralloc*, so the branch-and-bound procedure in *merge* can use the assigned sharing preference to merge registers.

3.3.1.4 An Example of *Ralloc*

The operation of *Ralloc* is illustrated by applying it to an example, named *ex2*, in Figure 3.10(a). The figure shows an SDFG taken from a part of the benchmark *DiffEq* [109], with the assumption that $V_I = \{u, dz, z, y\}$ and $V_O = \{u1\}$. Its lifetime

3.3. IMPLEMENTATION

table of variables is given in Figure 3.10(b).

In *Ralloc*, observability is enhanced first by calling BBREA with V_O and V_M as the two arguments. So a new register R_1 is allocated to the primary output $u1$ initially. Then e and f at cycle time 3 form the search space for R_1. To reduce the sequential depth as stated in SR2, a module sharing preference between $-_1$ and $-_2$ is assigned, which implies f, rather than e, should be assigned to R_1. The module sharing preference also implies another register sharing preference between f and c, and between e and u, to further save on the interconnection by making the operands assigned to the common registers. Notice that assigning a register sharing preference between e and u is also an attempt to enhance controllability of the register allocated to e. Then, at cycle time 2, c and d form the search space for R_1. Since in the previous cycle, c is preferred for sharing with f, R_1 is allocated to c. Next, at cycle time 1, a and b form the search space for R_1. If b is assigned to R_1, it implies that $*_1$ and $*_3$ should share the same multiplier to reduce the sequential depth to R_1 from the register allocated to a, which in turn implies that b is preferably shared with either u or dz to save on interconnection, which is not possible because dz has its lifetime overlapped with R_1 already, and u is preferred to share a register with e. On the other hand, if R_1 is allocated to a, the module sharing preference for $*_2$ and $*_3$, as well as the register sharing preference for b and z, is implied. So when BBREA returns, $\mathbf{R} = \{R_1\} = \{(a, c, f, u1)\}$, $V_{remain} = \{b, d, e\}$, and preferences are assigned to e and u, b and z, $-_1$ and $-_2$, and $*_2$ and $*_3$, respectively.

Controllability is enhanced next by calling BBLEA with V_I and V_{remain} as the two arguments. Each primary input is first assigned to a new temporary register: $tempR_1 = (z)$, $tempR_2 = (y)$, $tempR_3 = (u)$, and $tempR_4 = (dz)$. Since by BBREA, u and z have preferences assigned to e and b, respectively, we have $tempR_1 = (b, z)$ and $tempR_3 = (e, u)$. As for $tempR_2$, its search space at cycle time 2 consists of d only, so d is assigned to $tempR_2$. Furthermore, with this

72 CHAPTER 3. SEQUENTIAL DEPTH REDUCTION DURING ALLOCATION

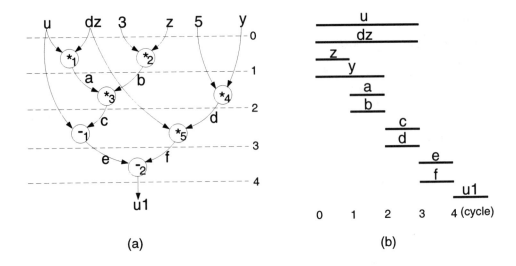

Figure 3.10: (a) SDFG of ex2; (b) its lifetime table.

assignment, a module sharing preference between $*_4$ and $*_5$ is assigned to save on interconnection. The search space for $tempR_4$ is empty, so no more variables can be assigned to it. Then, after $\mathbf{R}$ and $\mathbf{R}_{temp}$ are merged, we obtain the final register allocation with testability consideration as $\mathbf{R} = \{R_1, R_2, R_3, R_4, R_5\} = \{(a, c, f, u1), (b, z), (d, y), (e, u), (dz)\}$ with the minimum number of registers (i.e., 5).

With the above module sharing preferences, we can then easily perform module and interconnection allocations, which are discussed next, and obtain $\mathbf{M} = \{(-_{1,2}), (*_{1,4,5}), (*_{2,3})\}$. The DPCG for the allocation $\mathbf{R}$ and $\mathbf{M}$ is shown in Figure 3.11(a). We can see that the sequential depth from each register to the output register R_1 is at most 1. The fault coverage by STEED is 98.69% for a 2-bit implementation with a test generation time of 94.12 seconds.

3.3. IMPLEMENTATION

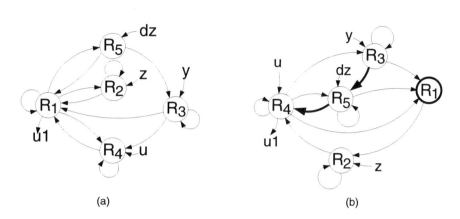

Figure 3.11: (a) DPCG of *ex2* allocated by PHITS; (b) DPCG of *ex2* allocated by the greedy LEA.

Another implementation of *ex2* using the same $\mathbf{M}$ but a different register allocation $\mathbf{R}' = \{R_1, R_2, R_3, R_4, R_5\} = \{(a), (b, c, z), (d, y), (e, u, u1), (dz, f)\}$ derived by the greedy left edge algorithm is also tested by STEED. The fault coverage for a 2-bit implementation is only 77.23% with a test generation time of 83.23 seconds. Its DPCG, shown in Figure 3.11(b), can demonstrate that the sequential depth from R_3 to R_4 is 2, and one register R_1 is neither directly controllable nor directly observable.

3.3.2 Module Allocation

After the module sharing preference is determined by register allocation, we can build a weighted module allocation graph to find the maximum-weight clique partitioning for module allocation [126], which is in general NP-complete. But suppose only a simple schedule and the regular module library are used. In such a case, Stok showed in [119] that the module allocation graph is indeed a comparability graph, and clique partitioning on such a graph can be found in polynomial time [51].

Notice that a **comparability graph** is a graph where if there exists an arc from node n_i to node n_j and an arc from node n_j to node n_k, $i \neq j \neq k$, then there must exist an arc from n_i to n_k. Stok proposed a simple procedure to transform the maximum-weight clique partitioning problem on a comparability graph into a classical maximum cost network flow problem [102]. The network flow problem can be solved exactly in polynomial time by a longest path algorithm.

A similar method is employed for module allocation in PHITS. We use the *build_up* procedure discussed in [102] with a modified version of the *Bellman-Ford* shortest path algorithm [38] to search for the longest path in the transformed network graph. The time complexity is $O(mne)$, where m is the number of modules allocated, and n and e are the numbers of nodes and arcs, respectively, in the transformed network graph.

3.3.3 Interconnection Allocation

There are two types of interconnection models commonly used, namely, bus and point-to-point. PHITS uses the point-to-point model with multiplexers between registers and modules. As discussed in [61], two sets of multiplexers are needed: one connecting a module's output to a register's input, and the other connecting a register's output to a module's input.

PHITS first allocates multiplexers to connect modules to registers. If two variables assigned to the same register are produced from the operations mapped to the same module, they can share the same input wire and thus a multiplexer input can be saved. So the input wires necessary for each register are determined first. Then a balanced binary tree structure of two-input multiplexers is formed to connect modules to the register.

To allocate multiplexers for register-to-module connection, a similar algorithm is used to minimize the number of multiplexers. Commutativity of operands is also considered for further minimization: If two additions $x + y$ and $x + z$ are mapped to one adder, then a multiplexer can be saved if y and z are assigned to the same input side of the adder, rather than assigning x and y, or x and z, to the same side.

3.4 Experimental Results

Five circuits were used to experiment with PHITS, where we assume all the variables born at the first cycle are primary inputs, the variables dying at the last cycle are primary outputs, and the others are intermediate variables. We first apply PHITS to the SDFG of each circuit for data path allocation, obtaining an output in *BLIF* format. The output is then optimized by SIS. STEED is then used to evaluate the testability. The experimental results are summarized in Table 3.1, where the column of #bit gives the bit-width of the data path, #mux in, the number of multiplexer inputs, %fc, the fault coverage, #abort/#fault, the ratio of the number of aborted faults to the total number of faults, and ATPG time, the CPU time used by STEED.

The first three circuits tested by PHITS are *ex2* (taken from the previous section), *oven-ctrl* (an oven controller example from [50]), and *real* (an example from [71]). The SDFG of *oven-ctrl* is depicted in Figure 3.12 with the assumption that $V_I = \{T1, T2, T3, T4, Tset\}$ and $V_O = \{Terror, Tdiff\}$. The SDFG of *real* is depicted in Figure 3.13 with the assumption that $V_I = \{a, b, c\}$ and $V_O = \{x1, x2\}$. Each of them is allocated by PHITS as well as a method based on the greedy left edge algorithm. The allocation details and testability comparisons are shown in Table 3.1. We can see that for each of the three examples, the 2-bit data path synthesized by PHITS has higher fault coverage than that synthesized by the greedy left edge algorithm, using a reasonable amount of test generation time and the same numbers of registers and modules.

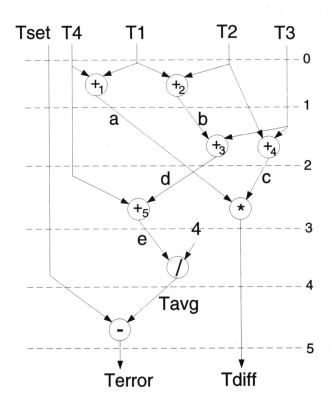

Figure 3.12: SDFG of *oven-ctrl*.

3.4. EXPERIMENTAL RESULTS

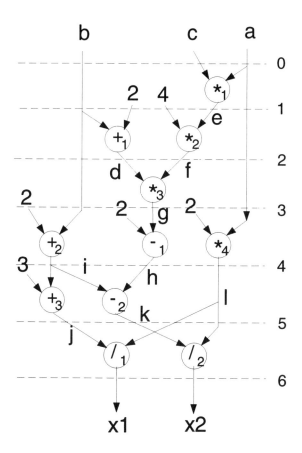

Figure 3.13: SDFG of *real*.

Table 3.1: Testability comparison by PHITS and the greedy LEA.

circuit (#bit)	allocation scheme	register allocation	module allocation	#mux in	%fc	#abort/ #fault	ATPG time
ex2 (2)	PHITS	$R_1 = (a, c, f, u1), R_2 = (b, z),$ $R_3 = (d, y), R_4 = (e, u),$ $R_5 = (dz)$	$(-_{1,2}),$ $(*_{1,4,5}),$ $(*_{2,3})$	15	98.69	3/305	94.12s
ex2 (2)	greedy LEA	$R_1 = (u, e, u1), R_2 = (b, c, z),$ $R_3 = (d, y), R_4 = (a),$ $R_5 = (dz, f)$	$(-_{1,2}),$ $(*_{1,4,5}),$ $(*_{2,3})$	16	77.23	68/303	83.23s
oven-ctrl (2)	PHITS	$R_1 = (T1, a), R_2 = (T2, c, Tdiff),$ $R_3 = (T3), R_4 = (T4),$ $R_5 = (b, d, e, Tavg, Terror),$ $R_6 = (Tset)$	$(+_{1,4}),$ $(+_{2,3,5}),$ $(-),$ $(*),(/)$	17	99.05	0/423	130.30s
oven-ctrl (2)	greedy LEA	$R_1 = (T1, a), R_2 = (T2, c),$ $R_3 = (T3, d, e, Tavg),$ $R_4 = (T4, Tdiff), R_5 = (b),$ $R_6 = (Tset, Terror)$	$(+_{1,4}),$ $(+_{2,3,5}),$ $(-),$ $(*),(/)$	20	93.98	25/432	665.3s
real (2)	PHITS	$R_1 = (a, l), R_2 = (b),$ $R_3 = (c, e, d, g, i, k, x1),$ $R_4 = (f, h, j, x2)$	$(*_{1,2,3,4}),$ $(+_{1,2,3}),$ $(-_{1,2}),$ $(/_1),(/_2)$	20	97.96	0/441	399.50s
real (2)	greedy LEA	$R_1 = (a, l, x1), R_2 = (b, i, j, x2),$ $R_3 = (c, e, f, g, h, k),$ $R_4 = (d)$	$(*_{1,2,3,4}),$ $(+_{1,2,3}),$ $(-_{1,2}),$ $(/_1),(/_2)$	16	90.55	21/381	55.10s
$DiffEq_A$ (2)	PHITS	$R_1 = (z, z1), R_2 = (a, c, f, u1),$ $R_3 = (b, d, g, y1),$ $R_4 = (A), R_5 = (dz), R_6 = (e, u),$ $R_7 = (y), R_8 = (ctrl)$	$(*_{1,4,6}),$ $(*_{2,3,5}),$ $(+_{1,2}),$ $(-_{1,2}),(<)$	18	99.78	0/448	709.0s
$Tseng_A$ (4)	PHITS	$R_1 = (v1, b, d, w3),$ $R_2 = (v4, c, e, w2),$ $R_3 = (v3, a),$ $R_4 = (v5, f), R_5 = (v2)$	$(*_{1,2}),$ $(+_{1,3}),$ $(+_2),$ $(-),(|)$	18	99.21	1/887	426.80s

3.4. EXPERIMENTAL RESULTS

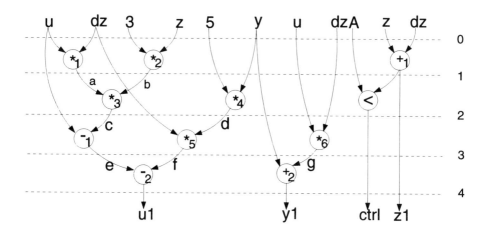

Figure 3.14: SDFG of $DiffEq_A$.

Figure 3.14 shows the SDFG of the fourth example $DiffEq_A$, taken from [109] with the assumption that $V_I = \{A, u, y, z, dz\}$ and $V_O = \{u1, y1, z1, ctrl\}$. The allocation details by PHITS are given in Table 3.1. Notice that register R_8 is just a 1-bit latch for the signal $ctrl$. Figure 3.15 depicts the DPCG of the derived architecture, where the sequential depth from each register to the nearest output register is at most one. The fault coverage by STEED for its 2-bit implementation is 99.78% in 709.0 seconds test generation time.

The fifth example, $Tseng_A$, is taken from [126]. The SDFG is shown in Figure 3.16(a) with the assumption that $V_I = \{v1, v2, v3, v4, v5\}$ and $V_O = \{w2, w3\}$. The allocation details by PHITS are also given in Table 3.1. Figure 3.16(b) shows the DPCG of its derived architecture, where the sequential depth from each register to the nearest output register is at most one. The fault coverage by STEED for its 4-bit implementation is 99.21% in test generation time of 426.8 seconds.

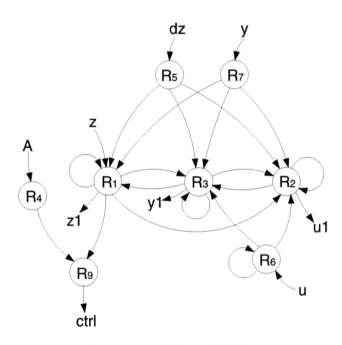

Figure 3.15: DPCG of *DiffEq$_A$*.

3.5. SUMMARY

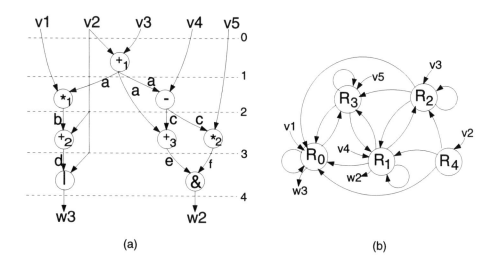

Figure 3.16: (a) SDFG of $Tseng_A$; (b) its DPCG.

3.5 Summary

In this chapter, we investigated the problem of high-level synthesis for testability in data path allocation assuming *a priori* no test strategy, and we proposed two synthesis rules for this purpose. Based on the two rules, a data path allocation scheme was implemented in PHITS, using a branch-and-bound left edge algorithm for register allocation with testability consideration. Several benchmarks were tested with PHITS and very high fault coverages were obtained in a reasonable amount of test generation time, using a minimal number of registers and modules.

Chapter 4

Sequential Loop Reduction During Allocation

In the previous chapter, two effective synthesis rules, SR1 and SR2, without any *a priori* test strategy assumption were proposed to synthesize testable simple data paths. For complex data paths where loop constructs are commonly used in the specifications, these rules may not be sufficient to produce highly testable architectures. This is because the cyclic data flows introduced by the loop constructs can create sequential loops in the final implementation, thus posing a major difficulty for sequential circuit testing [33].

In this chapter, we extend the results of the previous chapter to eliminate sequential loops during allocation. The allocation algorithm is actually differentiated into two levels of high-level test synthesis (HTS) considerations using non-scan and partial scan assumptions, respectively. The first-level HTS assumes no, or non-scan, test strategy, and can be economically applied to data paths with a small number of cyclic data flows. For larger data paths where a non-scan strategy cannot be sufficient, the second-level HTS uses the partial scan methodology and can take advantage of the available scan registers during high-level synthesis for both testability and area optimization. Experimental results on the benchmarks show that PHITS can synthesize highly testable data paths at low or no area overhead using the two-level HTS algorithm, when compared with other synthesis schemes and DFT methods.

84 CHAPTER 4. SEQUENTIAL LOOP REDUCTION DURING ALLOCATION

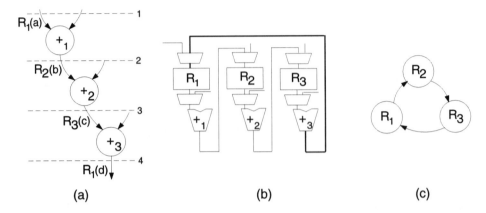

Figure 4.1: (a) Reuse of register R_1 for variables a and d in an acyclic SDFG; (b) feedback data path created in the architecture by reusing R_1; (c) corresponding sequential loop in the DPCG.

4.1 Effect of Sequential Loops on Testability

Since sequential loops in a circuit are mainly responsible for the difficulty of sequential test generation, we will study in this section how the sequential loop is created by allocation and how it affects testability, for acyclic and cyclic SDFGs, respectively.

4.1.1 Sequential Loops in Acyclic SDFG

Allocation can introduce sequential loops in the register-transfer architecture even when the initial SDFG is acyclic. A portion of an acyclic SDFG is shown in Figure 4.1(a), where $R_i(v_j)$ means that variable v_j is assigned to register R_i. We can see that R_1 has been used in the first cycle for storing one operand a of $+_1$, and reused in the fourth cycle for storing the result d of $+_3$. Figure 4.1(b) shows a fragment of its architecture, where a feedback data path is created by reusing R_1, provided that $+_1$, $+_2$, and $+_3$ are assigned to three different adders. A sequential loop in the

4.1. EFFECT OF SEQUENTIAL LOOPS ON TESTABILITY

DPCG corresponding to the feedback data path is shown in Figure 4.1(c).

Therefore, sequential loops can be created due to improper application of register sharing or module sharing by an allocation scheme. A concrete example is used next to illustrate the effect of sequential loops on testability for an acyclic SDFG, and show how to reduce sequential loops by proper resource sharing.

4.1.1.1 An Example

Figure 4.2 shows an acyclic SDFG, denoted as $ex3$, where the primary input set $V_I = \{x1, x4, x6, x7, x9\}$, the primary output set $V_O = \{y\}$, and the intermediate variable set $V_M = \{a, b, c, d, e, f, g\}$. The other inputs shown in the figure are constants. To show the effect of allocation in creating sequential loops, we will focus on the data flow in the SDFG from variables a to d, which is indicated by thick arcs in the figure. For the other data flows, the ideas discussed below can be applied as well. Suppose $ex3$ is synthesized by three different allocations $A1_{ex3}$, $A2_{ex3}$, and $A3_{ex3}$, with the corresponding DPCGs as shown in Figure 4.3(a), (b), and (c), respectively. The allocation results are summarized in Table 4.1, where the register allocation, the module allocation, and the number of multiplexer inputs (#mux in) are given. Notice that all three allocations use the same minimal number of registers and modules.

In the first allocation $A1_{ex3}$ shown in Table 4.1, the intermediate variables a, b, and c on the data flow of interest are each assigned to different registers (i.e., R_1, R_2, and R_3, respectively). Besides, the operations $+_2$ and $+_3$ in between those variables are assigned to different adders. Therefore, in its register-transfer architecture, there must be a sequential path corresponding to such an allocation: $R_1^a \xrightarrow{+2,6} R_2^b \xrightarrow{+3,7} R_3^c$. Furthermore, if operations $+_1$ and $+_4$ are assigned to the same adder, as in $A1_{ex3}$, this adder can store the result a in R_1 for $+_1$ and receive an operand c from R_3 for $+_4$, thus creating the sequential loop $R_1^a \xrightarrow{+2,6} R_2^b \xrightarrow{+3,7} R_3^c \xrightarrow{+1,4} R_1^a$. This sequential loop,

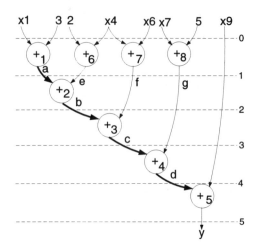

Figure 4.2: Example of an acyclic SDFG: ex3.

as reflected in its corresponding DPCG, is depicted by solid thick arcs in Figure 4.3(a). Notice that in Figure 4.3(a), $R_4 \to R_5 \to R_4$ is not a sequential loop because R_5 is an IO register.

For $A1_{ex3}$, we can also see from Figure 4.3(a) that module sharing of $+_1$ and $+_4$ also creates the sequential path $R_4^g \overset{+_{1,4}}{\to} R_1^a$, where $+_4$ takes an operand g from R_4 and $+_1$ stores the result a in R_1. This sequential path, together with $R_3^c \overset{+_{1,4}}{\to} R_4^d$ and the one discussed previously, forms a second sequential loop as indicated by dashed and solid thick arcs in the DPCG of Figure 4.3(a): $R_1 \to R_2 \to R_3 \to R_4 \to R_1$. Furthermore, the shared adder for $+_{1,4}$ also takes an operand $x1$ from R_1 for $+_1$ and stores the result d in R_4 for $+_4$, so there exists another sequential path $R_1^{x1} \overset{+_{1,4}}{\to} R_4^d$, introducing the third sequential loop, $R_1 \to R_4 \to R_1$, as indicated by dashed thick arcs in Figure 4.3(a). Therefore, $A1_{ex3}$ shows that sequential loops can be introduced unintentionally by module sharing, such as for $+_1$ and $+_4$ in this example. However, register sharing, as shown next for $A2_{ex3}$, can introduce sequential loops as well.

4.1. EFFECT OF SEQUENTIAL LOOPS ON TESTABILITY

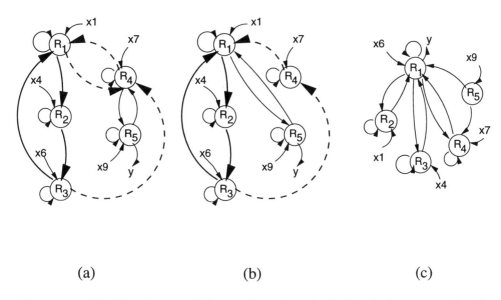

(a)　　　　　　　　　(b)　　　　　　　　　(c)

Figure 4.3: DPCGs of $ex3$ by different allocations (and their fault coverages): (a) $A1_{ex3}$ (84.27%); (b) $A2_{ex3}$ (58.43%); (c) $A3_{ex3}$ (100.00%).

Table 4.1: Effect of sequential loops on testability of $ex3$ by different allocations.

allocation scheme	register allocation	module allocation	#mux in	#loop	%fc	ATPG time
$A1_{ex3}$	$R_1 = (x1, a), R_2 = (x4, e, b),$ $R_3 = (x6, f, c),$ $R_4 = (x7, d, g),$ $R_5 = (x9, y)$	$(+_{1,4}),$ $(+_{2,6}),$ $(+_{3,7}),$ $(+_{5,8})$	19	3	84.27	159.3s
$A2_{ex3}$	$R_1 = (x1, a, d),$ $R_2 = (x4, e, b),$ $R_3 = (x6, f, c), R_4 = (x7, g),$ $R_5 = (x9, y)$	$(+_{1,5}),$ $(+_{2,6}),$ $(+_{3,7}),$ $(+_{4,8})$	17	2	58.43	213.5s
$A3_{ex3}$	$R_1 = (x6, e, b, c, d, y),$ $R_2 = (x1, a), R_3 = (x4, f),$ $R_4 = (x7, g), R_5 = (x9)$	$(+_{1,2}),$ $(+_{3,7}),(+_6)$ $(+_{4,5,8})$	16	0	100.00	26.1s

Suppose in the second allocation A2$_{ex3}$ given in Table 4.1, the two intermediate variables a and d are assigned to the same register R_1. Then a sequential loop for the data flow of interest can be created by the register sharing: $R_1^{a,d} \xrightarrow{+2,6} R_2^b \xrightarrow{+3,7} R_3^c \xrightarrow{+4,8} R_1^{a,d}$. This sequential loop is depicted by solid thick arcs in Figure 4.3(b). Moreover, there exist another two paths: $R_4^g \xrightarrow{+4,8} R_1^d$ where $+_4$ gets an operand g from R_4 and stores the result d in R_1, and $R_3^c \xrightarrow{+4,8} R_4^g$ where $+_4$ gets an operand c from R_3 and $+_8$ stores the result g in R_4. Those two paths, indicated by dashed thick arcs in its DPCG in Figure 4.3(b), can create another sequential loop for A2$_{ex3}$: $R_1 \rightarrow R_2 \rightarrow R_3 \rightarrow R_4 \rightarrow R_1$. The third allocation A3$_{ex3}$ given in Table 4.1, however, can synthesize $ex3$ without creating any sequential loops, as shown by its DPCG in Figure 4.3(c). The idea is basically to avoid creating sequential loops by introducing sequential paths to an IO register, such as R_1 in this case, during allocation.

The effects of sequential loops on the testability of $ex3$ by each allocation are also shown in Table 4.1, which gives the number of sequential loops (#loop), the fault coverage (%fc) by STEED on its 2-bit implementation, and the test generation time in seconds (ATPG time). We see that A3$_{ex3}$ can synthesize a completely testable data path using the shortest test generation time, while for the other two allocations, which can introduce sequential loops, the synthesized data paths have lower fault coverages in longer test generation time. Furthermore, A3$_{ex3}$ uses the fewest number of multiplexer inputs among the three, while requiring the same number of registers and modules. Thus, it is also the smallest of the three.

4.1.2 Sequential Loops in Cyclic SDFG

Cyclic data flows such as those imposed by *while* or *for* in the specification can, of course, also create sequential loops. With the loop construct, a boundary variable in the loop body, say v, will inherit its value from the previous loop iteration.

4.1. EFFECT OF SEQUENTIAL LOOPS ON TESTABILITY

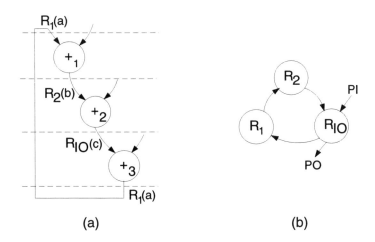

Figure 4.4: (a) Cyclic SDFG with boundary variable a assigned to R_1 and intermediate variable c to R_{IO}, an IO register; (b) its DPCG with the sequential loop broken by R_{IO}.

Usually, the register assignment to v, say R_v, is also invariant across loop iterations. Therefore, in the register-transfer architecture, there must exist a sequential loop starting from and ending at R_v, passing through some other registers allocated to the variables in the loop body that have direct or indirect data dependency on v.

Figure 4.4 (a) shows a cyclic SDFG where a is a boundary variable assigned to R_1. So a sequential loop is created by the cyclic data flow, as shown in its DPCG in Figure 4.4(b). To reduce the sequential loop, an IO register or a pre-determined scan register can be allocated to some variable on the cyclic data flow to effectively break the loop. In the above SDFG, variable c is assigned to an IO register to make the sequential loop much easier to control and observe. An example is used next to further discuss the above loop-breaking idea and the effect of sequential loops on testability.

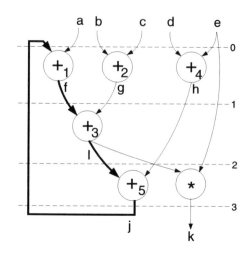

Figure 4.5: Example of a cyclic SDFG: $ex4$.

4.1.2.1 An Example

Figure 4.5 gives an example of a cyclic SDFG, denoted as $ex4$, where $V_I = \{a, b, c, d, e\}$, $V_O = \{k\}$, and $V_M = \{f, g, h, j, l\}$. Notice that the intermediate variable j is the boundary variable. This example is synthesized using three different allocation schemes $A1_{ex4}$, $A2_{ex4}$, and $A3_{ex4}$, with the corresponding DPCGs shown in Figure 4.6(a), (b), and (c), respectively. Table 4.2 lists the result for each allocation, and its effect on testability, where the column "max depth" gives the maximal sequential depth in the circuit. Also notice that all three allocations use the same minimal number of registers and modules.

In the first allocation $A1_{ex4}$, the three variables f, l, and j on the cyclic data flow in Figure 4.5 indicated by thick arcs are assigned to three different registers (i.e., R_1, R_2, and R_3, respectively). These three registers have a sequential path connected by the three addition operations $+_1$, $+_3$, and $+_5$, which are also assigned to three different adders by $A1_{ex4}$. Therefore, a sequential loop corresponding to such an allocation for the cyclic data flow is created: $R_1 \rightarrow R_2 \rightarrow R_3 \rightarrow R_1$, indicated by

4.1. EFFECT OF SEQUENTIAL LOOPS ON TESTABILITY

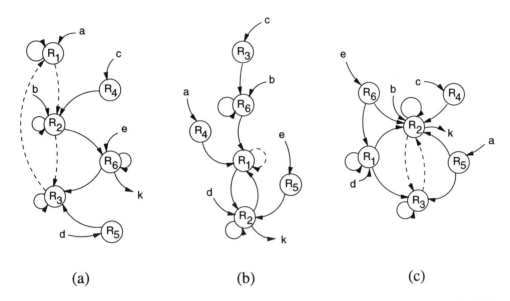

Figure 4.6: DPCGs of $ex4$ by different allocations (and their fault coverages): (a) $A1_{ex4}$ (33.79%); (b) $A2_{ex4}$ (67.66%); (c) $A3_{ex4}$ (99.43%).

Table 4.2: Effect of sequential loops on testability of $ex4$ by different allocations.

allocation scheme	register allocation	module allocation	#mux in	#loop	max depth	%fc	ATPG time
$A1_{ex4}$	$R_1=(a,f), R_2=(b,g,l),$ $R_3=(h,j), R_4=(c),$ $R_5=(d), R_6=(e,k)$	$(*),(+_1),$ $(+_{2,3}),$ $(+_{4,5})$	14	1	4	33.79	124.5s
$A2_{ex4}$	$R_1=(f,l,j), R_2=(d,h,k),$ $R_3=(c), R_4=(a),$ $R_5=(e), R_6=(b,g)$	$(*),(+_2),$ $(+_4),$ $(+_{1,3,5})$	10	0	3	67.66	86.5s
$A3_{ex4}$	$R_1=(d,h), R_2=(b,g,l,k),$ $R_3=(f,j), R_4=(c),$ $R_5=(a), R_6=(e)$	$(*),(+_2),$ $(+_4)$ $(+_{1,3,5})$	10	0	1	99.43	80.9s

92 CHAPTER 4. SEQUENTIAL LOOP REDUCTION DURING ALLOCATION

dashed arcs in its DPCG of Figure 4.6(a). The fault coverage for $A1_{ex4}$, as shown in Table 4.2, is only 33.79% for its 2-bit implementation. Notice that its maximal sequential depth 4 corresponds to the path $R_5 \to R_3 \to R_1 \to R_2 \to R_6$ in Figure 4.6(a).

As one possibility to improve testability, the above sequential loop introduced by the loop construct could be reduced in size. A straightforward idea is to reduce it to a trivial sequential loop by assigning all the variables on the cyclic data flow to the same register, since it has been shown that trivial sequential loops do not make test generation difficult [33]. Using this idea, the second allocation $A2_{ex4}$ in Table 4.2 assigns the three variables f, l, and j to the same register R_1. Furthermore, all the operations on the cyclic data flow are also assigned to the same adder $+_{1,3,5}$ by $A2_{ex4}$. As a result, the sequential loop imposed by the cyclic data flow is reduced to a trivial sequential loop, as indicated by the dashed arcs in Figure 4.6(b). Unfortunately, the fault coverage for its 2-bit implementation is still very low, though improved to 67.66%. This is because long sequential paths may be introduced as well, which can still give poor testability [81]. Actually, the register dedicated to the variables on the cyclic data flow, such as R_1 in this case, is neither directly controllable nor directly observable, and can be on a sequential path starting from an input register and ending at an output register, making the total sequential depth at least 2. In this example, we can see from its DPCG in Figure 4.6(b) that R_1 is on the sequential paths from R_4 to R_2 with depth 2, from R_6 to R_2 with depth 2, and from R_3 to R_2 with depth 3.

Allocation should avoid creating both sequential loops and long sequential paths. The third allocation $A3_{ex4}$ in Table 4.2 is thus proposed, where the essential idea is to break the sequential loop by allocating IO registers, such as R_2, to variables on the cyclic data flow, such as l, while at the same time reduce the lengths of the broken sequential loops by register allocation and module allocation. The corresponding

DPCG in Figure 4.6(c) shows that the sequential loop imposed by the cyclic data flow path, as indicated by the dashed arcs, is broken, and all the sequential paths from the input registers to the output register have lengths no longer than 2. The fault coverage for its 2-bit implementation, as shown in Table 4.2, is now improved to 99.43%, using the shortest test generation time. Besides, $A3_{ex4}$ also uses fewer multiplexer inputs than $A1_{ex4}$ and the same number as $A2_{ex4}$. Thus, it is also the smallest of the three.

From the previous discussion, we can therefore derive a new synthesis rule:

- SR3: Reduce sequential loops by
    - proper resource sharing to avoid creating sequential loops for acyclic DFGs, and
    - assign IO registers to break sequential loops in cyclic DFGs.

Since sequential depth may be unintentionally increased during sequential loop reduction, SR3 should work together with the synthesis rules SR1 and SR2 developed in the previous chapter to ensure testability.

4.2 Implementation

For data paths with few cyclic data flows, such as *ex4* mentioned in the previous section, we have shown that high testability can be easily obtained by HTS even without using any particular test strategy. To derive such testable architectures while minimizing area overhead, we propose our first-level HTS algorithm assuming no, or non-scan, test strategy for testability synthesis. On the other hand, for data paths with a large number of cyclic data flows, such as the *Elliptic Wave Filter* benchmark to be introduced, this approach is not sufficient. To complement our first-level testability synthesis method, we also propose the second-level HTS algorithm, assuming a less general but cost-efficient test methodology called partial scan, for those complex circuits. The second-level HTS algorithm attempts to derive an architecture best suited to partial scan during high-level synthesis itself, rather than add partial scan to the design later at the logic level [4]. This can substantially reduce the number of registers which need to be scanned.

The HTS scheme performs data path allocation on SDFGs in the following sequence: register allocation, module allocation, interconnection allocation. Among the three allocations, register allocation, where our testability synthesis is mainly embedded, is the most important step of the HTS scheme. Since the algorithms for module allocation and interconnection allocation are similar to the ones developed in Chapter 3, only register allocation is discussed next in detail.

4.2.1 Register Allocation

To achieve its goals, the register allocation should be able to assign variables on the cyclic data flow to an IO or scan register, or introduce sequential paths to an IO or scan register by appropriate register/module sharing. The previous allocations $A3_{ex3}$ and $A3_{ex4}$ actually used these ideas in their register allocations, and showed high testability could be obtained by them.

Figure 4.7 shows our register allocation procedure, called *RallocL*, using the above ideas. It can be applied to both non-scan and partial scan test strategies, and to both acyclic and cyclic SDFGs. Given the maximum number of registers that can be scanned, denoted as $\#scanR$, *RallocL* first attempts to select a set of $\#scanR$ or fewer boundary variables based on a heuristic discussed next, and assigns each boundary variable in the set to a scan register. This step can break the sequential loops caused by the cyclic data flow corresponding to each boundary variable in the set. Note that $\#scanR$ is usually small compared to the total number of registers, and can be set to 0 when no scan is allowed. Then *RallocL* allocates these scan

4.2. IMPLEMENTATION

registers, together with the normal IO registers, to other intermediate variables in order to further reduce the number of sequential loops and long sequential paths. For simplicity, we refer to the boundary variables selected for scan as *scan variables*, and the set of all scan variables as V_{scan}.

The heuristic we use to select V_{scan} from V_B prefers the boundary variable with a shorter lifetime. The idea is that if the scan register has only a small portion of its lifetime used by the scan variable, then *RallocL* can assign more intermediate variables to this scan register for more chances at loop reduction. Besides, the branch-and-bound versions of the left edge algorithm (BBLEA) and the right edge algorithm (BBREA) proposed in Chapter 3 are used as core subroutines in *RallocL*, which are effective in exploring the search space of intermediate variables for the best allocation to the IO or scan registers.

In Figure 4.7, *RallocL* takes an SDFG and $\#scanR$ as input parameters. It first selects the set of scan variables V_{scan} from V_B based on the heuristic discussed above (step 1). BBREA is then applied to assign the intermediate variables to the output or scan registers allocated to V_O or V_{scan} (step 2). Similarly, BBLEA is applied to assign the remaining intermediate variables to the input or scan registers allocated to V_I or V_{scan} (step 3). Then, if possible, the input registers and output registers are merged together for further saving in the number of registers (step 4). Here the procedure *merge* developed in Chapter 3 is used. The aim of these steps is to produce the best register allocation to reduce sequential loops as well as long sequential paths. If there are still intermediate variables unallocated, BBREA is applied to them by first forming registers with the boundary variables not included in V_{scan} (step 5), and then forming registers for the rest of the variables (step 6). Then all the allocated registers are returned when all the variables are allocated (step 7). Though *RallocL* does not guarantee register minimality, it does produce the minimal numbers of registers in almost all the cases, including the benchmarks

```
RallocL(#scanR) {
1.    V_scan = best_for_scan(V_B, #scanR);
2.    (R_output ∪ R_scan, V_remain) = BBREA(V_O ∪ V_scan, V_M);
3.    (R_input ∪ R_scan, V_remain) = BBLEA(V_I ∪ V_scan, V_remain);
4.    R_io = merge(R_input, R_output);
5.    (R_b, V_remain) = BBREA(V_B - V_scan, V_remain);
6.    if (V_remain ≠ ∅)
          R_m = BBREA(∅, V_remain);
7.    return(R_io ∪ R_scan ∪ R_b ∪ R_m);
}
```

Figure 4.7: Algorithm for *RallocL*.

considered later in the experimental result section.

4.2.2 An Example

We next take the SDFG of *ex4* in Figure 4.5 as a simple example to explain our register allocation. At step 1, the only boundary variable j is selected in V_{scan}, assuming only one scan register is given. At step 2, by branch-and-bound search for proper intermediate variable assignment, the set of scan registers $\mathbf{R}_{scan} = \{(f,j)\}$ and the set of output registers $\mathbf{R}_{output} = \{(g,l,k)\}$ are obtained for the scan variable j and the output variable k, respectively. At step 3, the set of input registers $\mathbf{R}_{input} = \{(a), (b), (c), (d,h), (e)\}$ can be similarly formed for all the primary inputs. Then at step 4, $\mathbf{R}_{input}$ and $\mathbf{R}_{output}$ can be merged into another set of registers $\mathbf{R}_{io} = \{(a), (b,g,l,k), (c), (d,h), (e)\}$. Since all the variables are assigned, $\mathbf{R}_{io}$ and $\mathbf{R}_{scan}$ are returned to the HTS algorithm.

4.3 Experimental Results

We have implemented the proposed two-level HTS algorithm in our PHITS system [79]. PHITS takes an SDFG and the number of scan registers (set to 0 if non-scan) as inputs, and produces a netlist output in *BLIF* format. We then apply SIS to the netlist to obtain optimized logic, and use STEED to evaluate the stuck-at fault testability of the 2-bit implementations.

Several benchmarks from the literature were synthesized by PHITS with very promising results. We will first present the results of our first-level HTS algorithm, assuming non-scan, and compare them with other synthesis schemes which do not have testability considerations. Then we will give the results of our second-level HTS algorithm, assuming partial scan, and compare them with the DFT approach where the scan registers are selected by the Lee-Reddy algorithm [72] after the architecture is synthesized by other schemes.

4.3.1 In the Non-Scan Environment

The benchmarks *Tseng*, *Paulin*, and *DiffEq* were synthesized by our first-level HTS algorithm in the non-scan environment, denoted by PHITS-NS. The experimental results for testability comparison between PHITS-NS and other algorithms are listed in Tables 4.3 and 4.4, where the column of #abort/#fault shows the ratio of the number of aborted faults to the total number of faults.

The SDFG of the first benchmark, *Tseng* from [9], is shown in Figure 4.8(a) with the assumption that $V_I = \{v4, v6, v10\}$, $V_O = \{v11\}$, and all the other variables are in V_M. We also assume that $V_B = \{v1, v2\}$. It was synthesized by PHITS-NS and three other allocation schemes collected in [9]: Splicer [97], Facet [126], and the one by Papachristou [99]. The results in Table 4.3 show that the architecture produced by PHITS-NS has the highest fault coverage in short test generation time

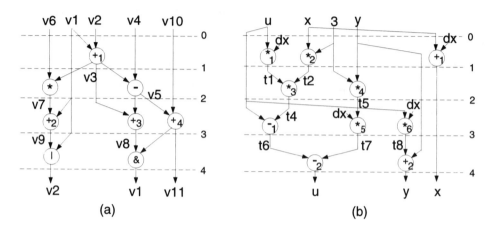

Figure 4.8: SDFGs of (a) *Tseng* and (b) *Paulin* used in [9].

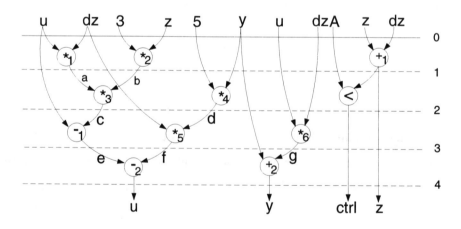

Figure 4.9: SDFG of *DiffEq*.

4.3. EXPERIMENTAL RESULTS

among the four. Besides, this testable architecture uses the fewest registers and the same minimal amount of modules with 11 multiplexer inputs. Notice that although fewer multiplexer inputs are required by the methods other than PHITS-NS, more registers are used by them.

Figure 4.9 depicts the SDFG of the second benchmark, *DiffEq* from [109], with the assumption that $V_I = \{u, dz, z, y, A\}$, $V_O = \{y, ctrl\}$, and all the other variables are in V_M. We also assume that $V_B = \{u, y, z\}$. PHITS-NS and three other allocations were applied to it: Splicer, HAL [109], and LYRA & ARYL [61]. The results in Table 4.3 show that PHITS-NS can produce the architecture with much higher fault coverage than all the other three in short test generation time. This architecture also uses the fewest registers and the same minimal amount of modules with 25 multiplexer inputs.

The SDFG of the third benchmark, *Paulin* from [9], is shown in Figure 4.8(b) with the assumption that $V_I = \{u, x, y\}$, $V_O = \{y\}$, and all the other variables are in V_M. We also assume that $V_B = \{u, y, x\}$. It was synthesized by PHITS-NS and two other allocation schemes, Splicer and HAL, collected in [9]. The results in Table 4.4 again show that the architecture produced by PHITS-NS has the highest fault coverage in short test generation time, and uses the fewest registers and the same minimal amount of modules with 24 multiplexer inputs.

4.3.2 In the Partial Scan Environment

Three benchmarks, *DiffEq*, *Paulin*, and *Elliptic Wave Filter* (*EWF*), were synthesized by our second-level HTS algorithm in the partial-scan environment, denoted by PHITS-PS.

For *EWF*, we assume that $V_I = \{IN\}$, $V_O = \{OUT\}$, and all the other variables are in V_M. The *EWF* benchmark [70] was first scheduled by the MPS scheduler [80], which assumes each module has unit delay, and will be introduced in Chapter 5. Its

Table 4.3: Testability comparison between PHITS-NS for non-scan and other synthesis methods without testability considerations (for benchmarks *Tseng* and *DiffEq*).

circuit	allocator	register allocation	module allocation	#mux in	%fc	#abort/ #fault	ATPG time
Tseng	PHITS-NS	$R_1 = (v10), R_2 = (v4, v7, v9)$, $R_3 = (v2, v3, v8), R_4 = (v1)$, $R_5 = (v6, v5, v11)$	$(+_{1,2}),(+_3)$, $(+_4),(-),(\&)$, $(|),(*)$	11	99.19	2/371	160.8s
	Splicer	$R_1 = (v10), R_2 = (v4)$, $R_3 = (v2, v3, v8), R_4 = (v1)$, $R_5 = (v7, v9), R_6 = (v6)$, $R_7 = (v5, v11)$	$(+_{1,2}),(+_3)$, $(+_4),(-),(\&)$, $(|),(*)$	8	96.95	18/388	2661.1s
	Facet	$R_1 = (v10), R_2 = (v4)$, $R_3 = (v2, v7, v9), R_4 = (v1)$, $R_5 = (v3, v8), R_6 = (v6)$, $R_7 = (v5, v11)$	$(+_{1,3}),(+_2)$, $(+_4),(-),(\&)$, $(|),(*)$	9	88.94	57/391	12407.4s
	Papachristou	$R_1 = (v10), R_2 = (v4)$, $R_3 = (v2), R_4 = (v1)$, $R_5 = (v7, v8), R_6 = (v6)$, $R_7 = (v5, v9), R_8 = (v3, v11)$	$(+_{1,4}),(+_2)$, $(+_3),(-),(\&)$, $(|),(*)$	8	86.02	55/392	9014.5s
DiffEq	PHITS-NS	$R_1 = (a, c, f), R_2 = (b, d, g)$, $R_3 = (u, e)$, $R_4 = (y), R_5 = (z)$, $R_6 = (A), R_7 = (dz)$, $R_8 = (ctrl)$	$(*_{1,4,5})$, $(*_{2,3,6})$, $(+_{1,2})$, $(-_{1,2})$, $(<)$	25	91.65	41/491	9574.3s
	Splicer	$R_1 = (a, d, g), R_2 = (b, c, f)$, $R_3 = (e), R_4 = (u)$, $R_5 = (y), R_6 = (z)$, $R_7 = (A), R_8 = (dz)$, $R_9 = (ctrl)$	$(*_{1,4,5})$, $(*_{2,3,5})$, $(+_{1,2})$, $(-_{1,2})$, $(<)$	27	79.89	109/542	28839.6s
	HAL	$R_1 = (a, c, f), R_2 = (b, d, g)$, $R_3 = (u, e)$, $R_4 = (y), R_5 = (z)$, $R_6 = (A), R_7 = (dz)$, $R_8 = (ctrl)$	$(*_{1,3,5})$, $(*_{2,3,5})$, $(+_{1,2})$, $(-_{1,2})$, $(<)$	20	73.13	126/454	803.6s
	LYRA & ARYL	$R_1 = (a, d, g), R_2 = (b, c, f)$, $R_3 = (u, e)$, $R_4 = (y), R_5 = (z)$, $R_6 = (A), R_7 = (dz)$, $R_8 = (ctrl)$	$(*_{1,4,6})$, $(*_{2,3,5})$, $(+_{1,2})$, $(-_{1,2})$, $(<)$	19	69.45	140/455	548.5s

4.3. EXPERIMENTAL RESULTS

Table 4.4: Testability comparison between PHITS-NS for non-scan and other synthesis methods without testability considerations (for benchmark *Paulin*).

circuit	allocator	register allocation	module allocation	#mux in	%fc	#abort/ #fault	ATPG time
Paulin	PHITS-NS	$R_1 = (t1, t4, t7), R_2 = (u, t6),$ $R_3 = (t2, t5, t8), R_4 = (y),$ $R_5 = (x), R_6 = (dx),$	$(*_{1,4,5}),$ $(*_{2,3,6}),$ $(+_{1,2}),(<)$ $(-_{1,2})$	24	91.78	41/450	17302.7s
	Splicer	$R_1 = (t1, t5, t8), R_2 = (t6),$ $R_3 = (t2, t4, t7), R_4 = (u),$ $R_5 = (y), R_6 = (x),$ $R_7 = (dx)$	$(*_{1,3,5}),$ $(*_{2,4,6}),$ $(+_{1,2}),(<)$ $(-_{1,2})$	26	77.73	116/485	27699.3s
	HAL	$R_1 = (t1, t5, t8), R_2 = (u, t6),$ $R_3 = (t2, t4, t7), R_4 = (y),$ $R_5 = (x), R_6 = (dx),$	$(*_{1,3,5}),$ $(*_{2,4,6}),$ $(+_{1,2}),(<)$ $(-_{1,2})$	20	69.02	127/397	438.4s

SDFG is shown in Figure 4.10 in textual representation where the execution cycle time is given at the beginning of each statement, and the variable with its name initial in upper case, such as $T2$, denotes the value in the previous loop iteration, while the variable with its name initial in lower case, such as $t2$, denotes the value in the current loop iteration. So two variables with names different only in the case of the first letter, such as $t2$ and $T2$, represent the different values of the same boundary variable in different iterations. For simplicity, each boundary variable is referred to only by its name with a lower case initial. Therefore, the set of all boundary variables V_B for *EWF* is $\{t2, t13, t18, t26, t38, t39, x15, k\}$. Table 4.5 lists the allocation statistics of the three benchmarks synthesized by PHITS-PS. The allocation statistics for *EWF* synthesized by PHITS-NS are also given in Table 4.5 for comparison purposes in the following experiment. For these allocations, PHITS-PS selects the boundary variables u in *DiffEq*, u in *Paulin*, and $t2, t13, t15$ in *EWF* as scan variables using the heuristic discussed in the previous section.

We will compare the testability results for circuits synthesized by PHITS-PS

CHAPTER 4. SEQUENTIAL LOOP REDUCTION DURING ALLOCATION

1:   $a = IN +_1 T2$;
1:   $t38 = T38 +_2 X15$;
2:   $b = T13 +_3 a$;
2:   $t33 = t38 +_4 K$;
3:   $x1 = b +_5 T26$;
3:   $g = t33 +_6 T39$;
4:   $e = x1 +_7 g$;
5:   $x2 = e *_1 2$;
6:   $d = b +_8 x2$;
7:   $x3 = e *_2 2$;
7:   $x4 = b +_9 d$;
7:   $x9 = d +_{10} e$;

8:   $f = x3 +_{11} g$;
8:   $x6 = x4 *_3 2$;
9:   $x5 = f +_{12} g$;
9:   $c = a +_{13} x6$;
10:   $x7 = x5 *_4 2$;
10:   $x8 = a +_{14} c$;
10:   $x12 = c +_{15} d$;
11:   $h = x7 +_{16} T39$;
11:   $x11 = x8 *_5 2$;
11:   $j = x12 +_{17} T18$;

12:   $x10 = f +_{18} h$;
12:   $x13 = T13 +_{19} h$;
12:   $x14 = IN +_{20} x11$;
12:   $x16 = j *_6 2$;
13:   $t26 = x9 +_{21} f$;
13:   $k = t38 +_{22} x10$;
13:   $OUT = x13 *_7 2$;
13:   $t18 = T18 +_{23} x16$;
14:   $t2 = x14 +_{24} c$;
14:   $x15 = k *_8 2$;
14:   $t39 = OUT +_{25} h$;
14:   $t13 = t18 +_{26} j$;

Figure 4.10: SDFG of *EWF* in textual representation.

Table 4.5: Allocation summary by PHITS in the partial scan and non-scan environments.

circuit	allocator	V_{scan}	register allocation	module allocation	#mux in
DiffEq	PHITS-PS	u	$R_1 = (A, d, f), R_2 = (a, c), R_3 = (u, e),$ $R_4 = (b), R_5 = (y), R_6 = (z), R_7 = (ctrl)$ $R_8 = (dz, g)$	$(*_{1,4,6}),(*_{2,3,5}),$ $(+_{1,2}),(-_{1,2}),(<)$	27
Paulin	PHITS-PS	u	$R_1 = (t1, t5, t8), R_2 = (t2, t4), R_3 = (u, t6),$ $R_4 = (dx, t7), R_5 = (y), R_6 = (x)$	$(*_{1,4,6}),(*_{2,3,5}),$ $(+_{1,2}),(-_{1,2}),(<)$	23
EWF	PHITS-PS	$t2,$ $t13,$ $x15$	$R_1 = (t2, x2, d, x7, h), R_2 = (t26, x3, f),$ $R_3 = (t13, x1, e, x4, x6, x12, j),$ $R_4 = (t33, g, x5, x8, x11, x13, OUT),$ $R_5 = (x15, a, x14), R_6 = (k, b, x16)$ $R_7 = (t18), R_8 = (t38), R_9 = (t39),$ $R_{10} = (IN, x10), R_{11} = (x9), R_{12} = (c)$	$(*_{1,2,3,4,5,6,7,8}),$ $(+_{1,3,5,7,8,9,11,}$ $+_{12,14,16,18,21,24}),$ $(+_{2,4,6,10,13,15},$ $+_{17,19,22,25}),$ $(+_{20,23,26})$	54
	PHITS-NS	$\emptyset$	$R_1 = (t2, g, c), R_2 = (t26, x3, f),$ $R_3 = (t13, x2, d, x7, j),$ $R_4 = (x15, t33, x1, e, x4, x6, x5, x12, h),$ $R_5 = (a, x8, x11, x13, OUT),$ $R_6 = (k, b, x10), R_7 = (t18), R_8 = (t38),$ $R_9 = (t39, x14), R_{10} = (IN, x16)$ $R_{11} = (x9)$	(same as the above)	50

4.3. EXPERIMENTAL RESULTS

and two DFT approaches. For the two DFT approaches, we first perform allocation on the benchmarks with no partial scan consideration to obtain the register-transfer architecture. Then in the first DFT approach, denoted by DFT-1, we apply the loop-breaking algorithm proposed by Lee and Reddy [72] on the derived DPCG to select a minimal or near-minimal set of n_1 scan registers such that the DPCG becomes acyclic.

Since n_1 may be greater than n_2, the number of scan registers required by PHITS-PS, it is also desirable to compare testability by scanning the same number of n_2 registers. Therefore, the second DFT method DFT-2 scans only the first n_2 out of the n_1 registers scanned by DFT-1. Note that the heuristic employed by the Lee-Reddy algorithm always selects the next scan register with the best cost for scan among the remaining registers until the DPCG becomes acyclic. So the n_2 registers scanned by DFT-2 have the best costs for scan. Also, since DFT-2 scans fewer registers than DFT-1, DFT-2 usually has lower testability than DFT-1. The testability comparison between PHITS-PS and the two DFT methods on each benchmark is shown in Table 4.6, which lists the numbers of registers (#reg) and multiplexer inputs (#mux in) derived from Tables 4.3, 4.4, and 4.5 for each allocation scheme, and the set of scan registers (R_{scan}) by PHITS-PS or the two DFT methods (DFT).

The first benchmark *Paulin* is synthesized by PHITS-PS, and the other two allocation schemes, Splicer and HAL, followed by DFT-1 and DFT-2. We can see from the results that PHITS-PS and Splicer followed by DFT-2 both scan one register and obtain 100% fault coverage, but PHITS-PS takes much shorter test generation time. As for HAL, neither DFT-1 nor DFT-2 can achieve 100% fault coverage.

The second benchmark *EWF* is synthesized by PHITS-PS and PHITS-NS followed by DFT-1 and DFT-2. The results show that the Lee-Reddy algorithm used

Table 4.6: Testability comparison between PHITS-PS and DFT methods.

circuit	allocator	DFT	#reg	#mux in	R_{scan}	$|R_{scan}|$	%fc	#abort/ #fault	ATPG time
Paulin	PHITS-PS	–	6	23	R_2	1	100.00	0/445	17.5s
	Splicer	DFT-1	7	26	R_5, R_4, R_6	3	100.00	0/487	226.6s
		DFT-2	7	26	R_5	1	100.00	0/487	1126.8s
	HAL	DFT-1	6	20	R_4, R_6	2	98.99	4/397	4.8s
		DFT-2	6	20	R_4	1	86.90	52/397	108.3s
EWF	PHITS-PS	–	12	54	R_1, R_4, R_8	3	97.46	24/944	140.4s
	PHITS-NS	DFT-1	11	50	$R_{10}, R_1, R_4,$ $R_3, R_5, R_6,$ R_9, R_7, R_{11}	9	98.48	9/856	66.0s
		DFT-2	11	50	R_{10}, R_1, R_4	3	77.35	172/852	590.8s
DiffEq	PHITS-PS	–	8	27	R_3	1	100.00	0/531	2619.2s
	Splicer	DFT-1	9	27	R_1, R_4	2	99.82	1/555	540.1s
		DFT-2	9	27	R_1	1	99.82	1/555	2523.6s
	HAL	DFT-1	8	20	R_2, R_1	2	99.34	3/454	12.0s
		DFT-2	8	20	R_2	1	86.78	60/454	279.7s
	LYRA & ARYL	DFT-1	9	19	R_1, R_3	2	98.67	6/452	21.0s
		DFT-2	9	19	R_1	1	83.63	74/452	325.5s

in DFT-1 needs to scan 9 out of 12 registers to make the DPCG acyclic and obtains 98.48% fault coverage. However, PHITS-PS needs only 3 scan registers to reach 97.46%, which is far better than 77.35% obtained by DFT-2 using the same number of scan registers.

The third benchmark *DiffEq* is synthesized by PHITS-PS, and three allocation schemes followed by DFT-1 and DFT-2: Splicer, HAL, and LYRA & ARYL. The results show that PHITS-PS with only one scan register can reach 100% fault coverage, while DFT-1 with two scan registers still has aborted faults for each of the three allocations. Furthermore, DFT-2, using the first scan register in DFT-1, has lower fault coverage for each of the three allocations than that for PHITS-PS, although both scan the same number of registers.

From the above results for the three benchmarks, we can see that PHITS-PS can scan fewer registers to achieve high fault coverage than DFT-1, and can give better testability than DFT-2 which scans the same number of registers.

4.4 Summary

In this chapter, we addressed the problem of how sequential loops introduced during high-level synthesis can impact sequential test generation. We proposed an HTS algorithm based on the reduction of sequential loops while minimizing area, and implemented it in the PHITS system. The proposed HTS algorithm considers two levels of testability synthesis: synthesis for non-scan (PHITS-NS), and synthesis for partial scan (PHITS-PS). The experimental results showed that (1) PHITS-NS could always synthesize a benchmark with the highest fault coverage in short test generation time, compared with other high-level synthesis algorithms that do not consider testability, and (2) PHITS-PS needed to scan fewer registers to achieve high fault coverage than the first DFT method which uses the Lee-Reddy algorithm, and could give better testability than the second DFT method which also scans the same number of registers.

Chapter 5

Testability Synthesis During Scheduling

The testability synthesis schemes presented in the previous chapters applied allocation procedures to DFGs that were already scheduled. However, scheduling should also take testability into consideration—otherwise the scheduler may choose a schedule that is good for other criteria but leads to inherently less testable architectures. Therefore, to be able to efficiently create a testable design, the design subspace imposed by a schedule should be restricted to mainly covering the design points where these allocation schemes are applicable.

In this chapter, we introduce a scheduling heuristic for testability based on the previous testability synthesis schemes SR1, SR2, and SR3 applied during data path allocation. We develop a *mobility path scheduling* algorithm in PHITS to implement this heuristic while also minimizing area. Experimental results on benchmark and example circuits show high fault coverage and short test generation time can be achieved with little or no area overhead.

5.1 Scheduling for Controllability/Observability Enhancement

Given an SDFG, the first synthesis rule SR1 attempts to enhance controllability and observability of a register by allocating it to a primary input or primary output. Therefore, for a partially allocated register R, the implementation of SR1 allocates R to a primary input/output v if the lifetime of v does not overlap with the current lifetime of R. However, due to the conflicting lifetimes defined by the scheduler, such a v may not exist.

Figure 5.1 gives a portion of an SDFG as an example with two different schedules, where $R(v_i)$ denotes the assignment of variable v_i to register R. Suppose register R is partially allocated to intermediate variables b, c, and other non-primary-input variables, and can only possibly use the primary input a for controllability enhancement according to SR1. In Figure 5.1(a), since the lifetime of a is $[0, 2]$ and it overlaps with that of R, the application of SR1 to R fails in this case. However, if operation $*$ is rescheduled to cycle 1 to decrease the lifetime of a to $[0, 1]$, as shown in Figure 5.1(b), a can now be assigned to R, which makes R directly controllable.

Another example to show the effect of scheduling on observability enhancement is depicted in Figure 5.2. Suppose register R_1 is partially allocated to intermediate variables w, x, and other non-primary-output variables, and R_2 is allocated to an intermediate variable y and a primary output z. Since R_2 is directly observable, one possibility to enhance the observability of R_1 is to merge R_2 with R_1. But for the schedule defined in Figure 5.2(a), such a merging is not possible because of conflicting lifetimes of R_1 and R_2. However, if the lifetime of R_2 can be decreased by scheduling operation / in cycle $t + 2$, R_2 can now be merged with R_1, as shown in Figure 5.2(b), thus making R_1 directly observable.

5.1. SCHEDULING FOR CONTROLLABILITY/OBSERVABILITY ENHANCEMENT

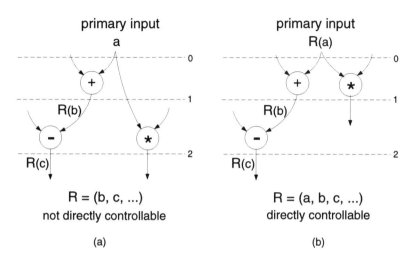

Figure 5.1: (a) R cannot be made directly controllable due to the conflicting lifetime of a; (b) by scheduling $*$ in cycle 1, SR1 can now be applied to R.

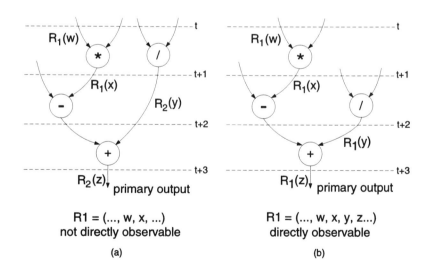

Figure 5.2: (a) R_1 cannot be made directly observable due to the conflicting lifetime of R_2; (b) by scheduling / in cycle $t+2$, SR1 can now by applied to R_1.

5.2 Scheduling for Sequential Depth/Loop Reduction

The other two allocation-for-testability rules SR2 and SR3 attempt to produce register or module sharing in such a way that sequential paths can be created to reduce sequential depths and loops. Again one potential difficulty is that such resource sharing can be prevented by a "bad" schedule.

For the SDFG example given in Figure 5.3(a), suppose register R_1 is allocated to variables w and x; R_2 is allocated to v, y, and z; and R_3 is allocated to s and u. Also suppose we want to reduce the sequential depth from R_1 to R_2. However, since R_1 and R_2 have overlapping lifetimes, and operations $*_1$ and $*_2$ have conflicting execution times, it is not possible to introduce a sequential path to reduce sequential depth by resource sharing. Therefore, in the register-transfer architecture, there must exist a sequential path $R_1^x \xrightarrow{-} R_3^u \xrightarrow{+} R_2^z$ corresponding to the data flow marked by thick arcs in Figure 5.3(a). Figure 5.3(b) depicts the DPCG with the sequential path of depth 2.

If the schedule in Figure 5.3(c) is used instead by rescheduling $*_2$ to $t + 2$, the sequential depth can be reduced by sharing $*_1$ and $*_2$. Although R_1 and R_2 still cannot be merged, the output of $*_1$ can now be directly observed at R_2 through the shared multiplier, which effectively makes R_1 easier to observe and hence reduces the sequential depth from R_1 to R_2. Figure 5.3(d) shows the DPCG with the reduced sequential depth.

The idea for reducing sequential loops by scheduling can be derived similarly.

Based on the above discussion, a new synthesis rule for scheduling can be summarized as:

- **SR4:** schedule operations to support the application of SR1, SR2, and SR3.

5.2. SCHEDULING FOR SEQUENTIAL DEPTH/LOOP REDUCTION 111

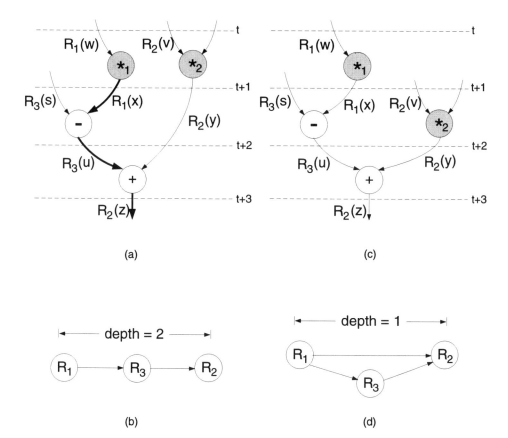

Figure 5.3: (a) Schedule where SR2 cannot be applied; (b) sequential depth is 2 from R_1 to R_2; (c) another schedule where SR2 can be applied; (d) sequential depth is reduced to 1.

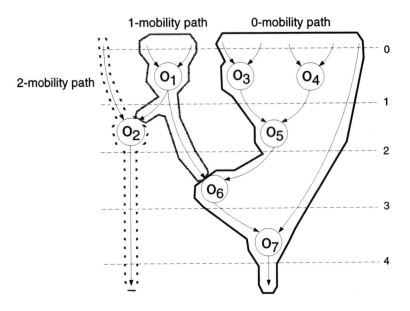

Figure 5.4: DFG partitioned into three mobility paths.

5.3 Implementation

Synthesis rule SR4 is applied using a new scheduling algorithm, called *mobility path scheduling* (MPS) [80], which is implemented in our PHITS system. MPS is an iterative/constructive scheduling algorithm that iteratively performs partial scheduling followed by *testMP*, a testability analysis procedure attempting to apply SR4.

We define a *k-mobility path* as a collection of data flows where the operations have the same mobility k. Hence the 0-mobility path is exactly the critical path. Figure 5.4 depicts an example of a DFG with 0-, 1-, and 2-mobility paths. Since each operation has its unique mobility, a DFG can be partitioned into several mobility paths with different k values.

5.3.1 Algorithm of MPS

The MPS algorithm first partitions the input DFG into several mobility paths where the operations on each path have the same mobility (i.e. the length of scheduling interval). Then, in each iteration, MPS selects the next unscheduled mobility path with the least mobility k, denoted as P_k, to perform partial scheduling. Hence, MPS is able to consider a sequence of data-dependent operations on the mobility path at the same time. Such a sequence, or the mobility path, has the property that all the operations on it are related by data dependency, and thus their mobilities are not totally independent during scheduling. For example, on the 1-mobility path in Figure 5.5, if + is first scheduled in cycle 2, the mobility of $*_2$ becomes 0 and only cycle 3 can be assigned. Therefore, scheduling more operations at a time that are related by the simple notion of mobility path can allow a large search space to produce a better schedule. Furthermore, MPS is also motivated by the observation that the allocator in PHITS prefers to assign data-dependent operations and variables to the same hardware for testability synthesis as well as potential savings on registers and interconnections.

After partial scheduling, testability analysis is performed on the scheduled mobility paths by *testMP*, a procedure that employs the same data path allocation algorithm as in PHITS, to indicate where the allocation-for-testability rules cannot be applied to the scheduled paths. Therefore, testability problems, such as bad controllability/observability or long sequential paths and loops, on the previously scheduled paths are examined so that they can be fixed by partial scheduling in the next iteration.

Partial scheduling can also consider load balancing to maximize resource utilization. That is, an operation on the selected mobility path is scheduled next if it has the least number of lightly loaded cycles in its slack. This will be further explained next.

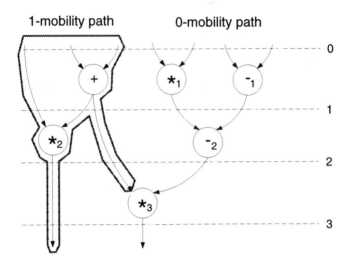

Figure 5.5: DFG partitioned into two mobility paths.

The above process of partial scheduling followed by testability analysis iterates until all the paths are scheduled. Since testability is taken into account at as early as the scheduling stage in the design flow, the final circuit can be synthesized with high testability.

The MPS algorithm, shown in Figure 5.6, first takes a DFG G as input and generates the slack/mobility for each operation (lines 1-3). Note that G is partitioned into a number of mobility paths at the same time. Then iterations are performed in a loop until all the operations in G are scheduled (line 4). In the loop body, the unscheduled mobility path P_k with the least mobility k is chosen for scheduling next (line 5). This is based on a similar heuristic in [98] where an operation with the smallest mobility is chosen to be scheduled next. Then partial scheduling (explained next), based on SR4 and load balancing, is performed on P_k (line 6). This step attempts to fix the testability problems in the previously scheduled mobility

5.3. IMPLEMENTATION

```
mobility_path_scheduling(G) {
1.    ASAP_scheduling(G);
2.    ALAP_scheduling(G);
3.    update_op_slack_and_mobility(G);
4.    while (unscheduled_op(G) ≠ ∅) {
5.        P_k = next_min_mobility_path(G);
6.        partial_scheduling(P_k, G);
7.        testMP(P_k, G); /* analyze testability on P_k */
      }
}
```

Figure 5.6: Mobility path scheduling algorithm.

paths by scheduling the operations on P_k with SR4 while considering load balancing. Then *testMP* is invoked to analyze the testability of the scheduled P_k based on SR4 (line 7).

The partial scheduling algorithm invoked above is described in Figure 5.7. It first assigns the execution time to an operation on P_k if its mobility becomes 0 (lines 1-3). Then the slack/mobility of each unscheduled operation is updated accordingly (line 4). Next, iterations are performed in a loop until all the operations on P_k are scheduled (line 5). Two major objectives are considered during scheduling P_k: testability and load balancing. In order to consider load balancing, the algorithm calculates the number of *light-load cycles* for each operation on P_k. A light-load cycle for an unscheduled operation o is a cycle in the slack of o which has a minimal number of scheduled operations of the same type. In Figure 5.5, for instance, the operation + with its slack [1, 2] has light-load cycles 1 and 2, while $*_2$ with its slack [2, 3] only has a light load cycle 2, with the assumption that the 0-mobility path is already scheduled. So in the loop body, the heuristic to pick the next operation to schedule is to select the one with the least light-load cycles (line 6). Among the light-load cycles, the one with the most preference for SR4 is assigned to this

```
partial_scheduling($P_k$, G) {
1.    foreach (operation o on $P_k$)
2.        if (o.earliest == o.latest) /* mobility becomes 0 */
3.            o.active = o.earliest; /* assign schedule */
4.    update_op_slack_and_mobility(G);
5.    while (unscheduled_op($P_k$) ≠ ∅) {
6.        (o, o.ll_cycles) = next_op_with_least_no_of_light_load_cycles($P_k$, G);
7.        o.active = most_preferred_cycle(o.ll_cycles, G);
      }
}
```

Figure 5.7: Partial scheduling algorithm.

operation (line 7).

5.4 Experimental Results

Three examples are used for experimenting with MPS in the PHITS system. Only a simple schedule is derived by MPS for each example (i.e. no operations execute over multiple cycles). Besides, we assume again that all the variables born at the first cycle in the SDFG are primary inputs, the variables dying at the last cycle are primary outputs, and the other variables are intermediate variables.

We first apply the MPS algorithm implemented in PHITS to the DFG of each example to obtain an SDFG with testability consideration. We then apply the data path allocator PHITS-NS developed in PHITS to the SDFG, assuming non-scan test strategy is used. The output is then optimized by SIS. Then STEED is used to evaluate the testability. For the purpose of comparison, each DFG is also scheduled by forced-directed scheduling (FDS) algorithm [107] without any testability consideration, followed by the same allocation algorithm PHITS-NS.

Since PHITS-NS can incorporate testability during allocation for a given SDFG,

5.4. EXPERIMENTAL RESULTS

the testability improvement by MPS given below could be much larger if compared with the circuits synthesized by FDS and other allocation schemes without testability consideration.

The experimental results are shown in Table 5.1, which gives the allocation details by PHITS-NS, the bit width of the data path (`#bit`), the fault coverage (`%fc`), the fault efficiency (`%fe`), the number of redundant faults (`#redun`), the number of aborted faults (`#abort`), the total number of faults (`#fault`), and the CPU time used by STEED (`ATPG time`).

The first example, denoted as ex, shows the effect of scheduling on testability in terms of fault coverage/efficiency and test generation time. Table 5.1 compares the testability of ex scheduled by MPS, shown in Figure 5.8, and by FDS, shown in Figure 5.9. Notice that FDS schedules the operation + in cycle 4 without increasing the number of modules.

We can see that for the 2-bit implementation, MPS can result in ex for which the test generation time is two orders of magnitude less than FDS, while it still has a high fault efficiency. For the 4-bit implementation, only on the circuit scheduled by MPS can STEED complete test generation before time-out (20 CPU hours), while also having very high fault coverage/efficiency. Note that the testability of ex scheduled by FDS can be much worse if it is not allocated by PHITS-NS. In addition, it is interesting to observe that the schedule derived by MPS requires five registers and has better testability than the schedule based on FDS which requires six registers.

The second example is the benchmark *DiffEq* [109]. Figure 5.10 depicts its SDFG after MPS determines the schedule. Table 5.1 also shows the experimental results for its 2-bit implementation after allocation and test generation. The fault efficiency is 100% and the test generation time is 709.0 seconds. *DiffEq* is also scheduled by FDS where the execution times of the operations $+_1$ and < in Figure

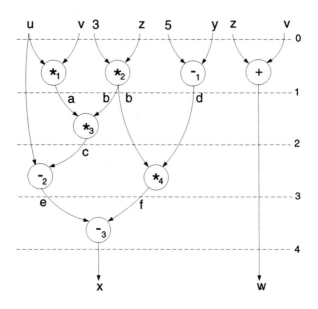

Figure 5.8: *ex* scheduled by MPS.

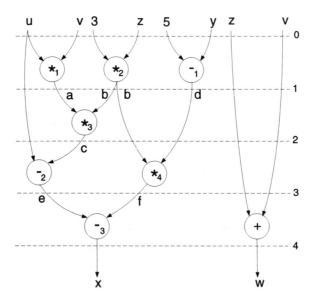

Figure 5.9: *ex* scheduled by FDS.

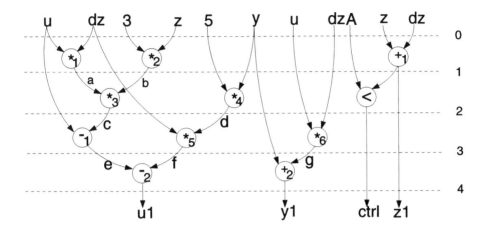

Figure 5.10: *DiffEq* scheduled by MPS.

5.10 are changed to cycles 3 and 4, respectively, without increasing the numbers of registers and modules. The results for its 2-bit implementation are shown in Table 5.1, with the same module allocation by PHITS-NS. The fault efficiency using FDS is still 100%, but the test generation is about 330 times slower when compared with the result for MPS. Note again that the testability by FDS can be much worse if *DiffEq* is not allocated by PHITS-NS.

The third example, denoted as *DCT*, is taken from a portion of an 8-point DCT signal flow graph [68]. Figures 5.11 and 5.12 show the SDFGs of *DCT* scheduled by MPS and FDS, respectively. The results given in Table 5.1 show that *DCT* scheduled by MPS can have 100% fault efficiency in shorter test generation time for both the 2- and 4-bit implementations. Notice again that the testability by FDS can be much worse if *DCT* is not allocated by PHITS-NS. Besides, we can also observe that the schedule determined by MPS requires five registers and has better testability than the schedule determined by FDS which requires six registers.

Table 5.1: Experimental results of testability by different schedules.

circuit	schedule	PHITS-NS register allocation	PHITS-NS module allocation	#mux in	#bit	%fc (%fe)	#redun/ #abort/ #fault	ATPG time
ex	MPS	$R_1=(y,d,f,x)$, $R_2=(w)$, $R_3=(u)$, $R_4=(v,a,c,e)$, $R_5=(z,b)$	$(*_{1,3})$, $(*_{2,4})$, $(+)$, $(-_{1,2,3})$	16	2	99.13 (99.71)	2/1/344	17.10s
					4	99.77 (100.00)	2/0/883	8.56h
	FDS	$R_1=(d,f,x)$, $R_2=(y,b)$, $R_3=(v)$, $R_4=(a,c,e,w)$, $R_5=(u)$, $R_6=(z)$	(same as above)	20	2	99.26 (99.26)	0/3/408	3926.20s
					4	(time-out at 20h; 522 out of 968 faults detected.)		
DiffEq	MPS	$R_1=(z,b,d,g,y1)$, $R_2=(z1)$, $R_3=(dz)$, $R_4=(a,c,f,u1)$, $R_5=(u,e)$, $R_6=(y)$	$(*_{1,4,6})$, $(*_{2,3,5})$, $(+_{1,2})$, $(-_{1,2})$, $(<)$	18	2	99.78 (100.00)	1/0/448	709.0s
	FDS	$R_1=(b,d,g,y1)$, $R_2=(z,z1)$, $R_3=(dz)$, $R_4=(a,c,f,u1)$, $R_5=(u,e)$, $R_6=(y)$	(same as above)	23	2	100.00 (100.00)	0/0/485	6.55h
DCT	MPS	$R_1=(p1,a,e,q4)$, $R_2=(p2,b,h,j,q2)$, $R_3=(p4,d,k)$, $R_4=(p3,g)$, $R_5=(c,f,q3)$	$(*_{1,3}),(*_{2,4})$, $(*_5)$, $(+_{1,3,4})$, $(+_{2,5}),(+_6)$, $(-_1),(-_2)$	24	2	99.78 (100.00)	1/0/463	144.9s
					4	100.00 (100.00)	0/0/993	2.02h
	FDS	$R_1=(p1,c,f,q3)$, $R_2=(p2)$, $R_3=(d,h,j,q2)$, $R_4=(p3,g)$, $R_5=(b,a,e,q4)$, $R_6=(p4,k)$,	(same as above)	25	2	100.00 (100.00)	0/0/497	1382.6s
					4	100.00 (100.00)	0/0/1029	18.62h

5.5. SUMMARY

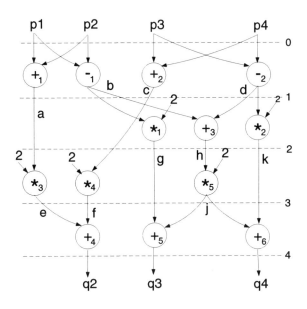

Figure 5.11: *DCT* scheduled by MPS.

5.5 Summary

In this chapter, we investigated the problem of high-level synthesis for testability in data path scheduling, and proposed a synthesis rule for this purpose, which extends the heuristics in previous chapters on allocation for testability. Based on this synthesis rule, a data path scheduling algorithm, called mobility path scheduling, was implemented in our PHITS system. Three examples were tested with PHITS and very high fault coverages/efficiencies were obtained in a relatively short amount of CPU time.

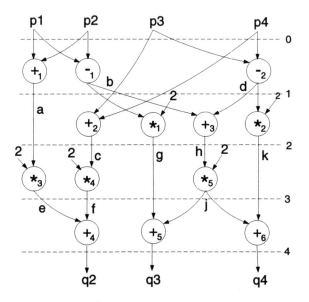

Figure 5.12: *DCT* scheduled by FDS.

Chapter 6

Conditional Resource Sharing for Testability

In high-level synthesis, resource sharing can be used to trade off design objectives during the design space exploration in order to obtain a good architecture [104]. Most resource sharing methods are usually divided into two types [63]: *unconditional resource sharing* and *conditional resource sharing*. Unconditional resource sharing, as used in those allocation algorithms developed in Chapters 3 and 4, is applied for sharing either functional modules for operations scheduled in different cycles or registers for variables without conflicting lifetimes. Conditional resource sharing performs, besides unconditional resource sharing, module and register sharing for operations and variables, respectively, at mutually exclusive parts of conditional branches.

Several previous works were proposed to perform conditional resource sharing for non-pipelined or pipelined data path synthesis, such as MAHA [104], Sehwa [103], Bridge [127], Cyber [132], Hwangs method [63], Kims method [65], and path-based scheduling [27]. However, these works focus on area or performance optimization. On the other hand, the high-level test synthesis (HTS) algorithms developed in the previous chapters do not consider conditional resource sharing.

In this chapter, we present a new conditional resource sharing method to synthesize highly testable non-pipelined data paths. This method is also implemented in

124 CHAPTER 6. CONDITIONAL RESOURCE SHARING FOR TESTABILITY

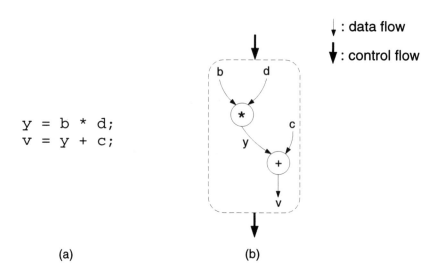

Figure 6.1: (a) Circuits behavior specified in a basic block; (b) its CDFG.

our PHITS system. A more general behavioral representation, called the *hierarchical control-data flow graph* (HCDFG), is used to model conditional branches. The conditional resource sharing method performs testability synthesis by traversing the HCDFG in a *postorder* fashion, with the assumption that either the non-scan or partial scan approach. Experimental results for the benchmarks show that our method, with *a priori* no test strategy assumption, can achieve higher fault coverage in shorter test generation time than an algorithm that disregards testability, and, with partial scan test assumption, can have high testability with fewer scan registers than some design-for-test methods.

6.1 Hierarchical Control-Data Flow Graph

The specification of circuit behavior contains both data flow and control flow information. The DFGs used in the previous chapters focus on modeling data flows

6.1. HIERARCHICAL CONTROL-DATA FLOW GRAPH

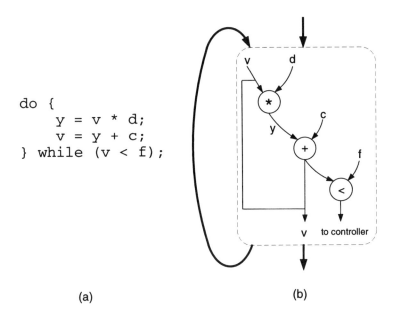

Figure 6.2: (a) Circuits behavior specified in a basic block with *while* at the end; (b) its CDFG.

for circuits that can be specified within a basic block. Such circuits usually have relatively simple control flows that can be easily derived from the DFGs. In Figures 6.1 and 6.2, two DFGs are shown with thick arcs indicating control flows between basic blocks. Such a DFG with explicit control flows can be called a *control-data flow graph* (CDFG).

The CDFG is very useful for modeling circuits with conditional branches in the behavioral specifications. Figure 6.3(a) gives an example of a behavioral specification with a conditional branch. The conditional branch defined by an if-then-else condition construct can specify two mutually exclusive paths of execution, called the *true block* (TB) and the *false block* (FB), arbitrated by an associated *condition*.

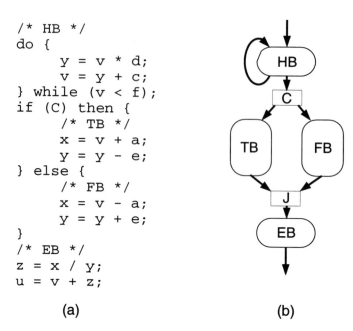

```
/* HB */
do {
    y = v * d;
    v = y + c;
} while (v < f);
if (C) then {
    /* TB */
    x = v + a;
    y = y - e;
} else {
    /* FB */
    x = v - a;
    y = y + e;
}
/* EB */
z = x / y;
u = v + z;
```

(a) (b)

Figure 6.3: (a) Example of a circuit description with a conditional branch; (b) layer representation for its CDFG.

The true block (false block) is defined as the block of statements selected for execution if the condition is true (false). Notice that for simplicity, we only consider two-case mutual exclusion, but it can be extended easily to handle multiple-case mutual exclusion specified by the switch construct.

Outside the conditional branch, we can define two other blocks: the *head block* (HB), for the block of statements scheduled for execution before the conditional branch, and the *end block* (EB), for the block of statements scheduled for execution after the conditional branch. Figure 6.3(b) shows a simplied CDFG corresponding to the specification in Figure 6.3(a), where nodes C and J represent, respectively, the starting and ending points of the conditional branch. We will call the structure of such a simplified CDFG, as in Figure 6.3(b), a **layer**. Moreover, a layer is said

6.1. HIERARCHICAL CONTROL-DATA FLOW GRAPH

to be **enabled** if the control flow activates one of its four blocks.

then it can also be represented by a layer structure identical to the one defining the true block or the false block itself. Such a recursive representation gives rise to our model of **hierarchical control-data flow graph** (HCDFG) for properly handling nested conditions commonly found in a circuit description. This is similar to the representations used in Olympus [89] and Hyper [113], but its formal definition is developed next for our proposed HTS algorithm which can take advantage of the formalism.

An HCDFG consists of a hierarchy of layers at different levels based on the nestedness of conditions, and can be defined recursively as follows:

- the **top layer** in an HCDFG is a layer that can be enabled without any conditions;

- a layer other than the top one is at a level below those layers whose C nodes correspond to the conditions nesting the layer, and can be enabled only under those conditions.

Figure 6.4(a) shows a "flattened" representation of an HCDFG example where the subscripts are used to uniquely identify a block, node, or layer. We can observe that the true block or the false block of a layer can be expanded into another layer one level below if it contains a condition construct. For instance, the false block of the top layer L_1 in Figure 6.4(a) has its C node for condition C_4, so this false block is expanded into another layer L_7 one level below. With this observation, a **layer tree** can be derived from an HCDFG where each node uniquely corresponds to a layer and an arc from node L_i to node L_j indicates that layer L_j is expanded from either the true block or the false block of layer L_i. Hence, in this definition, we can call L_i the **parent layer** of L_j, and L_j the **child layer** of L_i. Figure 6.4(b) shows the layer tree derived for the HCDFG in Figure 6.4(a). For simplicity, the left

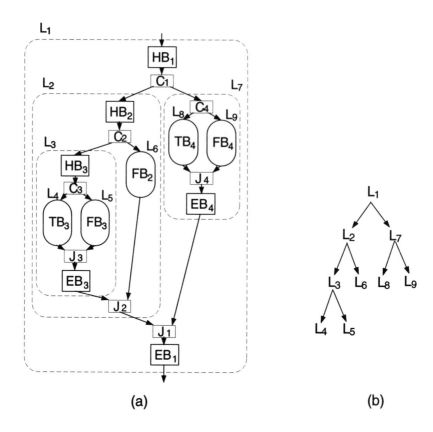

Figure 6.4: (a) Example of an HCDFG; (b) its layer tree.

(right) child layer always corresponds to the true block (false block) of its parent layer. The root node in a layer tree corresponds to the top layer, while each leaf node corresponds to a layer called the **bottom layer** containing only a head block.

If there is another condition construct nested inside a true block or a false block,

The lifetime table of variables can also handle conditional branches in an HCDFG by assigning a unique tag to the lifetimes of the variables in the same mutually exclusive block. Figure 6.5(a) shows an example of an acyclic HCDFG, denoted by $ex5$, with $V_I = \{a,b,d,e,g\}$, $V_O = \{h\}$, and $V_M = \{c,f\}$. Figure 6.5(b)

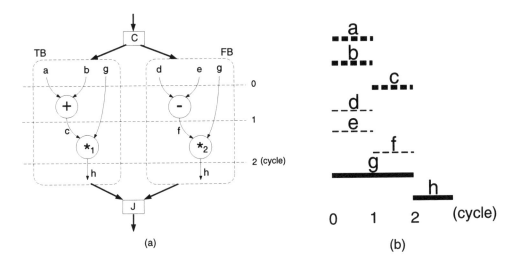

Figure 6.5: (a) Example of an acyclic HCDFG: ex5; (b) its lifetime table.

depicts its lifetime table with the thick solid lines for unconditional variables, and the thick and thin dashed lines for conditional variables in the true block and the false block, respectively. As proposed by Kurdahi and Parker in [71], register allocation for conditional variables can also be performed effectively on the lifetime table by the greedy left edge algorithm.

6.2 Effect of Conditional Resource Sharing

Conditional resource sharing explores the possibilities of sharing hardware components among the true blocks and the false blocks of conditional branches, in addition to the possibilities provided by unconditional resource sharing among disjoint cycles. Therefore, a conditional resource sharing method has a larger design space to evaluate, and may derive more than one valid design point under a given resource

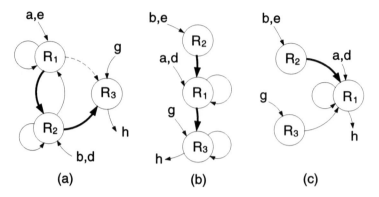

Figure 6.6: DPCGs of *ex5* by different allocations (and their fault coverages): (a) $A1_{ex5}$ (85.92%); (b) $A2_{ex5}$ (86.43%); (c) $A3_{ex5}$ (97.69%).

constraint. But not all these valid architectures are inherently testable. Hard-to-test circuit structures, such as those with long sequential paths and sequential loops, can be even more easily created due to the extra sharing possibilities allowed by mutual exclusion. Below we will demonstrate the effect of conditional resource sharing on testability based on different allocations used for two example circuits, *ex5* and *ex6*.

6.2.1 Conditional Resource Sharing for *ex5*

The first example *ex5* is an acyclic HCDFG shown in Figure 6.5(a). Suppose the data path is synthesized based on three different allocations $A1_{ex5}$, $A2_{ex5}$, and $A3_{ex5}$, with the derived DPCGs shown in Figure 6.6(a), (b), and (c), respectively. $A1_{ex5}$ and $A2_{ex5}$ are based on the greedy left edge algorithm, while $A3_{ex5}$ is based on our HTS algorithm to be proposed.

The allocation results are summarized in Table 6.1 where the register allocation, the module allocation, and the number of multiplexer inputs (`#mux in`) are given. Notice that all three allocations exploit the mutual exclusion of the two branches

6.2. EFFECT OF CONDITIONAL RESOURCE SHARING

and use the same minimal number of registers and modules. Below, we will first analyze the hard-to-test structures created by the partial allocations for the true block derived from $A1_{ex5}$, $A2_{ex5}$, and $A3_{ex5}$, and then look at the testability of the whole data path with conditional resource sharing between the true block and the false block.

In the first partial allocation of $A1_{ex5}$ for the true block, the variables a, c, and h are assigned to three different registers R_1, R_2, and R_3, respectively. Therefore, corresponding to the data flow $a \xrightarrow{\pm} c \xrightarrow{*1} h$ in the true block, there must be a sequential path of depth 2 in the register-transfer architecture, that is, $R_1^a \xrightarrow{\pm} R_2^c \xrightarrow{*1,2} R_3^h$, as indicated by the thick arcs in Figure 6.6(a). Similarly, for the data flow $b \xrightarrow{\pm} c \xrightarrow{*1} h$, the second partial allocation of $A2_{ex5}$ for the true block also has a corresponding sequential path of depth 2 in the register-transfer architecture, that is, $R_2^b \xrightarrow{\pm} R_1^c \xrightarrow{*1,2} R_3^h$, as indicated by the thick arcs in Figure 6.6(b). The third partial allocation of $A3_{ex5}$ for the true block, however, has the longest sequential path of length 1 only, that is, $R_2^b \rightarrow R_1^{c,h}$, for the data flow $a \xrightarrow{\pm} c \xrightarrow{*1} h$, as indicated by the thick arc in Figure 6.6(c). The partial allocations for the false block of $ex5$ can similarly be derived for $A1_{ex5}$, $A2_{ex5}$, and $A3_{ex5}$.

The complete data path of $ex5$ is synthesized based on the three allocations with conditional register sharing and module sharing between the true block and the false block. By $A1_{ex5}$, we can see from Figure 6.6(a) that the sequential depth from R_1 to R_3 is reduced from 2, when only the true block is considered, to 1 due to the new sequential path $R_1^f \xrightarrow{*1,2} R_3^h$, as indicated by the dashed arc in the DPCG, which is introduced by the data flow $f \xrightarrow{*3} h$ in the false block. But $A1_{ex5}$ can at the same time create a sequential loop by conditional register sharing, that is, $R_1^a \xrightarrow{\pm} R_2^{c,d} \xrightarrow{-} R_1^f$. This sequential loop corresponds to the combination of data flows $a \xrightarrow{\pm} c$ in the true block and $d \xrightarrow{-} f$ in the false block, and the register sharing for c in the true block and d in the false block. As for $A2_{ex5}$ and $A3_{ex5}$, conditional resource sharing

Table 6.1: Effect of conditional resource sharing on testability of *ex5* and *ex6* based on different allocations.

	register allocation	module allocation	#mux in	#loop	max depth	%fc	ATPG time
$A1_{ex5}$	$R_1 = (a, e, f)$, $R_2 = (b, c, d)$, $R_3 = (g, h)$	$(*_{1,2})$, $(+)$, $(-)$	8	1	1	85.92	300.6 s
$A2_{ex5}$	$R_1 = (a, c, d, f)$, $R_2 = (b, e)$, $R_3 = (g, h)$	$(*_{1,2})$, $(+)$, $(-)$	5	0	2	86.43	182.6s
$A3_{ex5}$	$R_1 = (a, c, d, f, h)$, $R_2 = (b, e)$, $R_3 = (g)$	$(*_{1,2})$, $(+)$, $(-)$	4	0	1	97.69	347.7s
$A1_{ex6}$	$R_1 = (m), R_2 = (b, h)$, $R_3 = (c, j), R_4 = (d, f, k)$, $R_5 = (e, l, g)$	$(+_{2,3,4})$, $(+_1),(*)$, $(-)$	10	2	4	64.54	28.4s
$A2_{ex6}$	$R_1 = (h, j, k, m)$, $R_2 = (b), R_3 = (c)$, $R_4 = (d, f), R_5 = (e, l, g)$	$(+_{1,3})$, $(+_{2,4})$, $(-),(*)$	9	0	1	98.37	124.9s

does not introduce any sequential loops or longer sequential paths than the ones created by the partial allocation for the true block. Hence, the maximum sequential depths for $A2_{ex5}$ and $A3_{ex5}$ are still 2 and 1, respectively.

The testability of each implementation is also shown in Table 6.1, which gives the number of sequential loops (#loop), the maximum sequential depth (max depth), the stuck-at fault coverage (%fc) by STEED on its 4-bit implementation, and the test generation time in seconds (ATPG time). The same backtrack limit was used in each case. Since $A3_{ex5}$ does not create sequential loops or long sequential paths, it can synthesize a highly testable data path for *ex5* with the fault coverage about 11% better than that for $A1_{ex5}$ or $A2_{ex5}$. Furthermore, $A3_{ex5}$ uses the fewest number of multiplexer inputs among the three, while requiring the same number of registers and modules. Thus, it is also the smallest of the three.

6.2. EFFECT OF CONDITIONAL RESOURCE SHARING

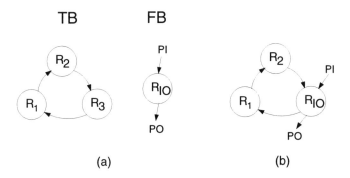

Figure 6.7: (a) A sequential loop in the true block and an IO register in the false block; (b) the sequential loop is broken by sharing the IO register.

We can observe from this example that in order to produce highly testable data paths for acyclic HCDFGs, the synthesis algorithm, while optimizing area, should at the same time:

- eliminate sequential loops and reduce sequential depths individually for a true block and a false block;

- not create sequential loops or long sequential paths for the whole data path by conditional resource sharing between a true block and a false block;

- if possible, further break the existing sequential loops or long sequential depths by conditional resource sharing.

Figure 6.7 gives an example where a sequential loop created in the true block can be broken by an IO register allocated in the false block if the IO register is shared by the two mutually exclusive blocks.

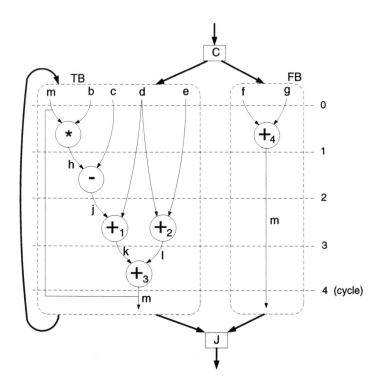

Figure 6.8: Example of a cyclic HCDFG: *ex6*.

6.2. EFFECT OF CONDITIONAL RESOURCE SHARING

6.2.2 Conditional Resource Sharing for $ex6$

Figure 6.8 shows the second example, denoted by $ex6$, of a cyclic HCDFG with $V_I = \{b,c,d,e,f,g\}$, $V_O = \{m\}$, $V_M = \{h,j,k,l\}$, and $V_B = \{m\}$ induced by the cyclic data flow $m \xrightarrow{*} h \to j \xrightarrow{+1} k \xrightarrow{+3} m$ in the true block. Two allocations, $A1_{ex6}$ and $A2_{ex6}$, are obtained to synthesize the data path for $ex6$ with the same minimal numbers of registers and modules, as also summarized in Table 6.1. Notice that $A1_{ex6}$ is based on the greedy left edge algorithm, and $A2_{ex6}$ is based on the HTS algorithm to be proposed.

Since the partial allocation of $A1_{ex6}$ for the true block assigns each variable on the cyclic data flow to a different register, a sequential loop is created, that is, $R_1^m \xrightarrow{*} R_2^h \to R_3^j \xrightarrow{+1} R_4^k \xrightarrow{+3} R_1^m$, as indicated by the thick arcs in Figure 6.9(a). The partial allocation of $A2_{ex6}$ for the true block, on the other hand, can reduce the cyclic data flow into a trivial sequential loop, as indicated by the thick arc in Figure 6.9(b), by assigning all the variables on the cyclic flow to a single register R_1.

For the whole data path of $ex6$, conditional resource sharing by $A1_{ex6}$ introduces the second sequential loop, $R_1 \to R_2 \to R_3 \to R_4 \to R_5 \to R_1$, and a sequential path of length 4, $R_2 \to R_3 \to R_4 \to R_5 \to R_1$. By $A2_{ex6}$, however, no sequential loop is created and the maximum sequential depth is only 1. Therefore, the testability listed in Table 6.1 for the 2-bit implementation of $ex6$ shows that $A2_{ex6}$ can produce a highly testable data path with the fault coverage being about 34% better than that for $A1_{ex6}$, again given the same backtrack limit. Furthermore, since $A2_{ex6}$ uses fewer multiplexer inputs than $A1_{ex6}$, $A2_{ex6}$ is also smaller than $A1_{ex6}$.

From $ex6$, one more observation can be made for synthesizing highly testable data paths with cyclic HCDFGs, in addition to the previous ones made for acyclic HCDFGs. That is, while optimizing area, the synthesis algorithm should at the same time:

136 CHAPTER 6. CONDITIONAL RESOURCE SHARING FOR TESTABILITY

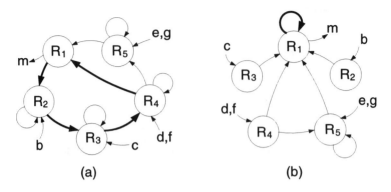

Figure 6.9: DPCGs of *ex6* by different allocations (and their fault coverages): (a) $A1_{ex6}$ (64.54%); (b) $A2_{ex6}$ (98.37%).

- break or reduce in size the sequential loops imposed by the loop constructs in the synthesis of a true block or a false block individually, or by conditional resource sharing between a true block and a false block.

6.3 A Conditional Resource Sharing Method for HTS

Based on the previous discussions, we propose a conditional resource sharing method to reduce sequential depths/loops starting from HCDFGs with conditional branches, while optimizing area. The proposed method is mainly embedded in the register allocation algorithm. After register allocation, module allocation and interconnection allocation are then performed to complete the allocation, based on the algorithms developed in Chapter 3. We will describe the algorithm of register allocation in detail. We will first focus on register allocation without assuming any test strategy beforehand. We refer to this case as the non-scan case. Then we will discuss its extension to the case where partial scan may be required.

6.3. A CONDITIONAL RESOURCE SHARING METHOD FOR HTS

6.3.1 Register Allocation

To effectively perform testability synthesis during register allocation, we develop a register allocation algorithm that recursively traverses the layer tree derived from an HCDFG in a *postorder* fashion, and allocates registers for each layer with testability consideration while optimizing area. In this algorithm, the measures of controllability from primary inputs and observability from primary outputs at the top layer are passed to each child layer from the parent layer while traversing down the layer tree. Similarly, testability synthesized at each child layer using the controllability/observability measures is attempted to be preserved and further enforced, if possible, at the parent layer while ascending up the layer tree.

We next describe the register allocation algorithm *RallocB* for testability synthesis in a head block or an end block, and present the register allocation algorithm *Traversal* for a layer to preserve or enforce testability synthesized by *RallocB* at the lower layers.

6.3.2 Allocation for an HB or EB: *RallocB*

The lifetimes of variables in a head block or an end block of a layer can be easily identified from the lifetime table of an HCDFG. So a left edge or right edge algorithm for register allocation can be employed for the portion of the lifetime table corresponding to the variables in a head block or an end block. Though Kurdahi's greedy left edge algorithm [71] can be applied to obtain the minimal number of registers, it does not consider testability. Therefore, the branch-and-bound versions of left edge algorithm (BBLEA) and right edge algorithm (BBREA) developed in Chapter 3 for sequential depth/loop reduction are used as core subroutines in our register allocation algorithm, called *RallocB*, for a head block and an end block with modifications for layer tree traversal.

```
RallocB_obsr(B) {
    (R_1, V_rest) = BBREA(B.V_out, B.V_rest);
    (R_2, V_rest) = BBLEA(B.V_in, V_rest);
    R = merge(R, Ralloc_min(V_rest) ∪
                 merge(R_1, R_2));
}

RallocB_ctrl(B) {
    (R_1, V_rest) = BBLEA(B.V_in, B.V_rest);
    (R_2, V_rest) = BBREA(B.V_out, V_rest);
    R = merge(R, Ralloc_min(V_rest) ∪
                 merge(R_1, R_2));
}
```

Figure 6.10: Algorithms for *RallocB$_{obsr}$* and *RallocB$_{ctrl}$*.

Before allocating registers for a head block or an end block of a layer, *RallocB* can deduce, from the partial allocation done in the layers at higher levels, the input/output registers partially allocated to primary inputs/outputs and the sequential paths between the partially allocated registers. Therefore, with such a controllability/observability measure, *RallocB* can apply BBLEA and BBREA to explore the search space of intermediate variables for testability synthesis. Preference for module sharing between two operations is also set during the branch-and-bound searching process for further reduction in sequential depth/loop as well as area saving in interconnection. The module sharing preference can also be used later for module allocation.

Figure 6.10 describes two types of *RallocB* algorithms for register allocation of a head block or an end block: *RallocB$_{obsr}$*, which enhances observability first, and *RallocB$_{ctrl}$*, which enhances controllability first. *RallocB$_{obsr}$* takes as an input a head block or an end block, denoted by B, with its variables partitioned into three subsets: the variables fanning into B ($B.V_{in}$), the variables fanning out of B ($B.V_{out}$), and the remaining variables in B ($B.V_{rest}$). *RallocB$_{obsr}$* first applies

BBREA to assign as many variables in $B.V_{rest}$ as possible to the registers that are allocated to the variables in $B.V_{out}$ in order to enhance observability, based on the observability measure passed from the higher layers. BBREA can return the set of registers it allocates in $\mathbf{R}_1$ and the set of the remaining unallocated variables of $B.V_{rest}$ in V_{rest}. Then $RallocB_{obsr}$ applies BBLEA to assign as many variables as possible in V_{rest} to the registers that are allocated to the variables in $B.V_{in}$ in order to enhance controllability, based on the controllability measure passed from the higher layers. BBLEA can return the set of registers it allocates in $\mathbf{R}_2$ and the set of the remaining unallocated variables in V_{rest}.

If V_{rest} is not null, it is then assigned by $Ralloc_{min}$, developed in Chapter 3 for obtaining a minimal number of registers. Since some registers in $\mathbf{R}_1$ and $\mathbf{R}_2$ do not have conflicting lifetimes, they can be merged together by the function *merge* developed in Chapter 3. Finally, $RallocB_{obsr}$ applies *merge* again to merge all the registers allocated locally for the variables in B with the registers previously allocated for other blocks to form a new set of registers in $\mathbf{R}$.

Similarly, $RallocB_{ctrl}$ listed in Figure 6.10 can apply BBLEA before BBREA to enhance controllability first.

6.3.3 Allocation for a Layer: *Traversal*

The register allocation for a layer is performed by a *postorder* traversal of the layer tree. That is, a parent layer must have its two children layers finish their register allocations before it can finish. Therefore, based on the above *RallocB* algorithm, two types of register allocation algorithms for a layer are presented in Figure 6.11: $Traversal_{obsr}$, which enhances observability first, and $Traversal_{ctrl}$, which enhances controllability first.

Given a layer L as an input, $Traversal_{obsr}$ performs $RallocB_{obsr}$ on the end block first and on the head block last to proceed in the backward direction in L for

```
Traversal_obsr(L) {
    RallocB_obsr(L.EB);
    R_T = Traversal_obsr(L.TB);
    R_F = Traversal_obsr(L.FB);
    overlay(R_T,R_F);
    RallocB_obsr(L.HB);
}


Traversal_ctrl(L) {
    RallocB_ctrl(L.HB);
    R_T = Traversal_ctrl(L.TB);
    R_F = Traversal_ctrl(L.FB);
    overlay(R_T,R_F);
    RallocB_ctrl(L.EB);
}
```

Figure 6.11: Algorithms for $Traversal_{obsr}$ and $Traversal_{ctrl}$.

observability enhancement. $Traversal_{ctrl}$, on the other hand, performs $RallocB_{ctrl}$ on the head block first and on the end block last to proceed in the forward direction for controllability enhancement. As for the true block and the false block of L, each algorithm recursively calls itself to perform register allocation for the true block before the false block. Notice that the above recursion actually corresponds to a postorder traversal of a layer tree rooted at L. Then the two mutually exclusive register sets derived for the true block and the false block are combined by the function *overlay*, which attempts to merge two registers from each set if no new sequential loop or long sequential path will be created, or the existing loop/paths, if any, can be broken.

6.3. A CONDITIONAL RESOURCE SHARING METHOD FOR HTS

```
RallocC(HCDFG) {
    Traversal_obsr(L_top);
    Traversal_ctrl(L_top);
    Ralloc_min(V_rest);
}
```

Figure 6.12: Algorithm for *RallocC*.

6.3.4 *RallocC* for Complete Allocation

The complete register allocation *RallocC*, as given in Figure 6.12, first applies $Traversal_{obsr}$ to proceed in the backward direction in the top layer of an HCDFG, denoted as L_{top}, for observability enhancement. $Traversal_{ctrl}$ is then applied to proceed in the forward direction in L_{top} to allocate registers for the unallocated variables in order to enhance controllability. If there are still unallocated variables, $RallocC_{min}$ is called to allocate registers.

The *RallocC* algorithm can handle both acyclic and cyclic HCDFGs, but it does not guarantee minimality of registers. Experimental results given in the next section, however, show that it allocated a minimal number of registers for the benchmarks we synthesized.

6.3.5 *RallocC* in the Partial Scan Environment

Since circuits with a large number of cyclic data flows can create complex sequential loops, a partial scan test strategy may become necessary for synthesizing highly testable data paths. *RallocC* can be easily extended to support partial scan for cyclic HCDFG with conditional branches by selecting a small number of boundary variables with the shortest lifetimes to form the set of scan registers, using the same heuristic developed in Chapter 4.

6.4 Experimental Results

We also implemented the proposed conditional resource sharing method in our PHITS system [78]. PHITS takes an HCDFG and the number of scan registers (set to 0 if non-scan) as inputs and produces a netlist output in *BLIF* format. We then apply SIS to the netlist to obtain the optimized logic, and use STEED to evaluate the stuck-at fault testability of the 2-bit implementations.

For each benchmark, the HCDFG is flattened into a single layer for simplicity, with the variables born at cycle 0 put in V_I, the variables dying at the last cycle put in V_O, and the other variables put in V_M. Since we focus on synthesis of the data path, it is assumed that the conditional branching at each C node is determined by a synthesized controller which can, based on the status signals from the data path in the current cycle, send back the control signals to the data path in the next cycle.

PHITS-NS is used to denote the application of PHITS under the non-scan assumption, and is compared with the greedy LEA [71] which does not have testability consideration. PHITS-PS is used to denote the application of PHITS under the partial scan assumption, and is compared with other synthesis schemes without testability consideration followed by two logic-level DFT approaches, DFT-1 and DFT-2, defined in Chapter 4. For DFT-1 and DFT-2, the scan registers are determined by the Lee-Reddy algorithm [72] after the architecture is synthesized by the greedy left edge algorithm. DFT-1 scans all the n_1 registers selected by the Lee-Reddy algorithm, while DFT-2 scans only the first n_2 out of the n_1 registers from DFT-1, where n_2 is the number of registers scanned by PHITS-PS for the benchmark.

6.4. EXPERIMENTAL RESULTS

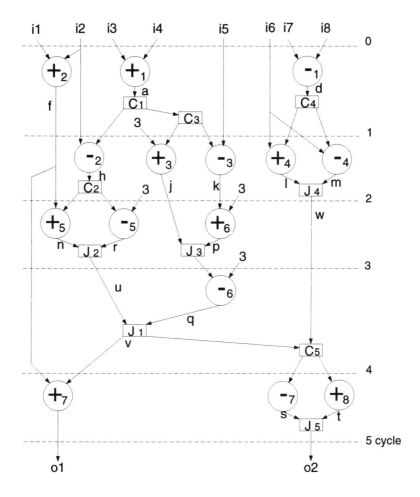

Figure 6.13: HCDFG of *Sehwa*.

6.4.1 In the Non-Scan Environment

The acyclic HCDFGs of the three benchmarks, *Sehwa* [103], *Maha* [104], and *Kim* [65], are depicted in Figures 6.13, 6.14, and 6.15, respectively. Each data path is synthesized by both PHITS-NS and the greedy left edge algorithm, with the results shown in Table 6.2. We can see that all three architectures produced by PHITS-NS have higher fault coverages in shorter test generation times than those produced by the greedy left edge algorithm. Besides, these three testable architectures require the same minimal number of registers and modules as required by the greedy left edge algorithm.

To show the effectiveness of our approach on cyclic HCDFGs, we also apply PHITS-NS and the greedy left edge algorithm to two examples of cyclic HCDFGs for testability comparison. The first example of a cyclic HCDFG, $Sehwa_L$, is created by making o_2 of *Sehwa* an intermediate variable and feeding it back to i_3. Notice that $Sehwa_L$ can also be allocated with the same allocation for *Sehwa* by the greedy left edge algorithm, except that R_1 becomes only an input register. The second example of a cyclic HCDFG, $Maha_L$, is made by feeding variable r of *Maha* back to i_1. $Maha_L$ can also be allocated with the allocation used for *Maha* by the greedy left edge algorithm without any change in the data path. The results in Table 6.3 show that PHITS-NS can produce testable architectures for the two cyclic HCDFGs with much higher fault coverage in shorter test generation time than the greedy left edge algorithm. Besides, PHITS-NS requires the same minimal number of registers and modules as required by the greedy left edge algorithm.

6.4.2 In the Partial Scan Environment

As we can see above for $Maha_L$, although its data path synthesized by PHITS-NS is more testable than the one by the greedy left edge algorithm, its fault coverage is not

6.4. EXPERIMENTAL RESULTS

Table 6.2: Testability comparisons for the acyclic benchmarks synthesized by PHITS-NS and the greedy LEA.

circuit	test synthesis	register allocation	module allocation	#mux in	%fc	#abort/ #fault	ATPG time
Sehwa	PHITS-NS	$R_1 = (o1, v, u, q, n, r, h,$ $a, j, k, i3), R_2 = (o2, s,$ $t, w, l, m, d, i7),$ $R_3 = (f, i1), R_4 = (i2),$ $R_5 = (i4), R_6 = (i5),$ $R_7 = (i7), R_8 = (i8)$	$(+_{2,3,5,6,7}),$ $(+_{1,4,8}),$ $(-_{2,3,5,6}),$ $(-_{1,4,7})$	22	99.44	1/534	316.8s
	greedy LEA	$R_1 = (o2, s, t, v, u, q, n, r,$ $k, j, h, a, i3),$ $R_2 = (o1, f, i1),$ $R_3 = (l, i2), R_4 = (i5),$ $R_5 = (d, i7), R_6 = (i8),$ $R_7 = (i4), R_8 = (i6)$	same as the above	28	92.17	46/600	4745.7s
Maha	PHITS-NS	$R_1 = (o, s, t, r, k, e, f, c, q,$ $p, l, n, m, g, d, a, i2),$ $R_2 = (u, j, h, b, i5),$ $R_3 = (i1), R_4 = (i3),$ $R_5 = (i4), R_6 = (i6)$	$(+_{1,2,4,5,6,7}),$ $(+_{3,8}),$ $(-_{2,3,5,6,7}),$ $(-_{1,4,8})$	23	94.04	26/534	867.0s
	greedy LEA	$R_1 = (o, s, t, i2),$ $R_2 = (r, q, k, e, f, c, i1),$ $R_3 = (p, l, n, m, g, d, i4),$ $R_4 = (o, t, s, u, h, j, b, i5),$ $R_5 = (i3), R_6 = (i6)$	same as the above	21	77.28	117/515	10986.9s
Kim	PHITS-NS	$R_1 = (o1, y, z, i, v, x, u,$ $n, e, q, f, b, s, h, c, i1),$ $R_2 = (w, t, j, i5, r, g, i4,$ $p, a, i3),$ $R_3 = (m, i2),$ $R_4 = (o2, k),$ $R_5 = (l, d, i6)$	$(+_{1,5,7,9,10,12,}$ $_{13,14,15,16}),$ $(+_{2,3,4,6,8,11}),$ $(-_{1,2,4,5,7,8}),$ $(-_{3,6,9})$	35	99.84	1/608	5450.7s
	greedy LEA	$R_1 = (o1, y, z, i, v, x, p, a,$ $i3, r, g, b, i4, w, t, j, c, i5),$ $R_2 = (o2, k, i6),$ $R_3 = (l, d, i1),$ $R_4 = (s, q, u, n, i2),$ $R_5 = (e, m, f, h)$	$(+_{1,5,7,9,12,14,}$ $_{15,16}),$ $(+_{2,3,4,6,8,10,}$ $_{11}),$ $(-_{1,2,4,5,8,9}),$ $(-_{3,6,7})$	39	97.55	14/572	19898.8s

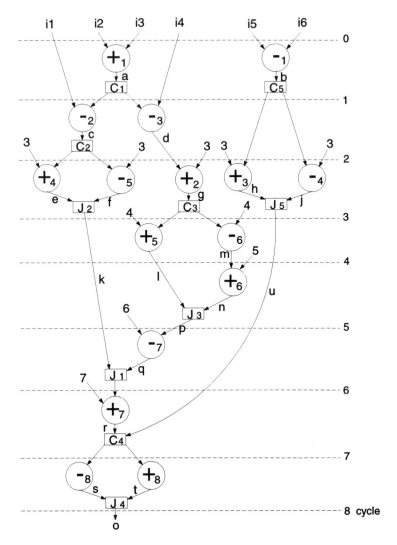

Figure 6.14: HCDFG of *Maha*.

6.4. EXPERIMENTAL RESULTS

Table 6.3: Testability comparisons for the cyclic benchmarks synthesized by PHITS-NS and the greedy LEA.

circuit	test synthesis	register allocation	module allocation	#mux in	%fc	#abort/ #fault	ATPG time
$Sehwa_L$	PHITS-NS	$R_1 = (o1, v, u, q, n, r, h, a, j, k, i4), R_2 = (o2, s, t, w, l, m, d, i3), R_3 = (f, i1), R_4 = (i2), R_5 = (i5), R_6 = (i6), R_7 = (i7), R_8 = (i8)$	$(+_{2,3,5,6,7}), (+_{1,4,8}), (-_{2,3,5,6}), (-_{1,4,7})$	22	97.61	12/544	826.5s
	greedy LEA	same as by greedy LEA for $Sehwa$, except R_1 is only an input register		28	86.50	80/600	7610.7s
$Maha_L$	PHITS-NS	$R_1 = (r, k, e, f, q, c, p, l, n, m, g, d, i1), R_2 = (o, t, s, u, h, j, b, i5), R_3 = (o, s, t, i2), R_4 = (i3), R_5 = (i4), R_6 = (i6)$	$(+_{1,2,4,5,6,7}), (+_{3,8}), (-_{2,3,5,6,7}), (-_{1,4,8})$	23	92.36	36/471	1784.2s
	greedy LEA	same as by greedy LEA for $Maha$		21	77.28	117/515	10986.9s

148 CHAPTER 6. CONDITIONAL RESOURCE SHARING FOR TESTABILITY

sufficiently high. Therefore, we apply PHITS-PS to the cyclic examples, exploiting partial scan during synthesis itself, and compare them in terms of testability with the architectures derived by the greedy left edge algorithm followed by DFT-1 or DFT-2. An additional example of a cyclic HCDFG, Kim_L, is created for comparison by making o_2 of Kim an intermediate variable and feeding it back to i_1.

The results are shown in Table 6.4. For each example, PHITS-PS scans only the register that is allocated to the boundary variable. When comparing PHITS-PS and DFT-1, we can see that PHITS-PS can select fewer scan registers than DFT-1, yet can still give a very high fault coverage. Moreover, in the case of $Sehwa_L$, DFT-1 scans one more register than PHITS-PS, yet gives a lower fault coverage in longer test generation time. As for PHITS-PS and DFT-2 which both scan the same number of registers, we can see that PHITS-PS can always achieve higher fault coverage in shorter test generation time for the three examples than DFT-2.

6.5 Summary

In this chapter, we addressed the problem of how conditional resource sharing allowed by mutual exclusion during high-level synthesis can impact sequential test generation. We proposed a conditional resource sharing method for highly testable data path synthesis while minimizing area, and implemented it in our PHITS system. The experimental results showed that (1) assuming non-scan strategy, PHITS could always synthesize a benchmark with higher fault coverage in shorter test generation time than the greedy left edge algorithm that does not consider testability, while (2) assuming partial-scan test strategy, PHITS needed to scan fewer registers than the DFT-1 method, which uses the Lee-Reddy algorithm, and still achieved very high fault coverage; it could also give better testability in shorter test generation time than the DFT-2 method which scans the same number of registers.

6.5. SUMMARY

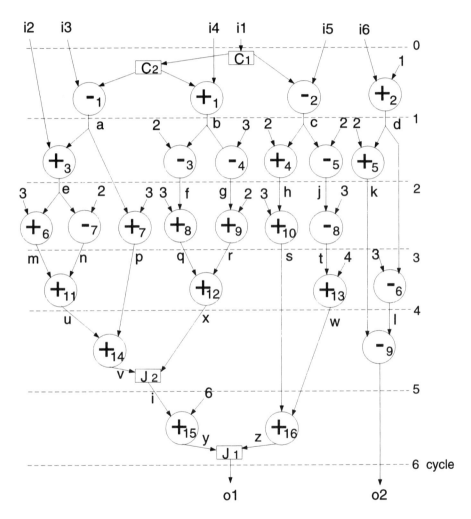

Figure 6.15: HCDFG of *Kim*.

Table 6.4: Testability comparisons for cyclic HCDFGs by PHITS-PS and DFT methods.

circuit	test synthesis	allocation	R_{scan}	%fc	#abort/ #fault	ATPG time
$Sehwa_L$	PHITS-PS	same as by PHITS-NS for $Sehwa_L$	R_2	99.82	0/600	244.5s
	DFT-1	same as by greedy LEA for $Sehwa_L$	R_1,R_3	97.36	15/600	600.9s
	DFT-2	same as by greedy LEA for $Sehwa_L$	R_1	97.20	16/607	1418.2s
$Maha_L$	PHITS-PS	same as by PHITS-NS for $Maha_L$	R_1	99.58	2/471	135.4s
	DFT-1	same as by greedy LEA for $Maha_L$	R_2,R_1	100.00	0/518	85.6s
	DFT-2	same as by greedy LEA for $Maha_L$	R_2	99.23	4/518	430.0s
Kim_L	PHITS-PS	same as by PHITS-NS for Kim except $i1$ is assigned to R_4, and #mux in = 41	R_1	99.50	3/602	1754.0s
	DFT-1	same as by greedy LEA for Kim except $i1$ is assigned to R_2, $i6$ is assigned to R_3, and #mux in = 39	R_4,R_3,R_5	100.00	0/572	467.3s
	DFT-2	same as the above	R_4	96.77	17/572	2013.2s

Chapter 7

State-of-the-Art High-Level Test Synthesis

There has been increasing research interest in high-level test synthesis. To provide a comprehensive understanding of this area, although it is impossible to make an exhaustive survey, this chapter gives an in-depth overview of several representative current works on high-level test synthesis, in addition to the PHITS system discussed in the previous chapters. Test strategies of built-in self test and scan are presented, respectively, in Sections 7.1 and 7.2, while RTL test synthesis is presented in Section 7.3. Other related works are briefly surveyed in Section 7.4. The summary is given in 7.5.

7.1 Synthesis for Built-In Self Test

The growth of chip complexity has been reducing the direct controllability and observability of internal signals through IO pins. Furthermore, the increasing cost of using automatic test equipment makes external testing very expensive. Therefore, built-in self test (BIST), which builds testing circuitry into the design to test itself without using an external test equipment to access test data through the limited number of IO pins, is gaining popularity by testing industry. In this section, three high-level BIST synthesis systems are discussed: SYNTEST in Section 7.1.1, RALLOC in Section 7.1.2, and SYNCBIST in Section 7.1.3.

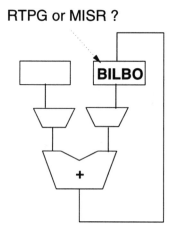

Figure 7.1: BILBO of a self-adjacent register cannot perform RTPG and MISR at the same time.

The common BIST structure used by the high-level BIST synthesis is the BILBO register [67], which can perform normal register, serial scan in/out, random test pattern generation (RTPG), and multiple-input signature register (MISR), as explained in Section 2.7.1. So, in a data path design, all the registers could be converted to BILBO registers. However, such simple conversion does not guarantee BIST testability because, as in the case of Figure 7.1, the BILBO register cannot be both RTPG and MISR at the same time to test a module whose input and output share the same register, referred to as self-adjacent register. To cope with the problem with self-adjacent registers, SYNTEST and SYNCBIST propose methods to completely avoid producing such self-adjacent registers during high-level synthesis, while RAL-LOC uses CBILBO registers [133] for self-adjacent registers after the occurrence of self-adjacent registers is minimized. The CBILBO register is able to perform both RTPG and MISR at the same time. In addition, SYNTEST exploits the circuits functionality to further reduce the overhead introduced by BILBO registers, while SYNCBIST also attempts to maximize test concurrency.

7.1. SYNTHESIS FOR BUILT-IN SELF TEST

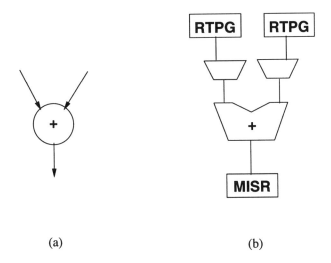

Figure 7.2: An operation in the SDFG of (a) is mapped to the testable functional block (TFB) in (b).

7.1.1 SYNTEST: Case Western Reserve University

SYNTEST [56] is a high-level test synthesis system developed by Harmanani, Papachristou, Chiu, and Nourani at Case Western Reserve University. It applies BIST methodology to produce testable data path designs by the following three major steps during resource allocation, with the assumption that scheduling is done:

- step 1: Avoid register self-adjacency entirely to guarantee that the design is structurally testable by BILBO.

Such structural testability can be achieved by mapping each operation in an SDFG to a *testable functional block* (TFB), as shown in Figure 7.2, with the constraint that the output of the TFB cannot drive any of the same TFBs inputs. This constraint guarantees the resulting data path has no self-adjacent registers, and is therefore self-testable. However, this initial testable design is produced with no resource sharing consideration.

- step 2: Merge compatible resources to minimize area and delay.

  Stepwise refinement is performed on the initial testable design to minimize area and delay by merging compatible resources to improve utilization. Two resources are compatible if they cause no conflict and introduce no self-loop. So two compatible TFBs can be merged. Cost functions are used to measure gains by such resource sharing to guide the refinement.

  As a result, no self-adjacent registers are created for the optimized design. Each module is still directly controllable by the RTPGs and directly observable by the MISR. This type of BIST design is called *maximal BIST strategy*, whose sequential depth is 1.

- step 3: Further exploit functional testability of modules to reduce BIST overhead without impacting the fault coverage too much [37].

  During TFB merging process of each refinement in step 2, functional testability of each module can be used to remove unnecessary RTPG or MISR function to reduce overhead. Two major functional testability metrics are considered here for pseudorandom BIST:

  - *randomness*: an entropy-based approach to measure randomness at a modules output.

    Entropy is a characterization of a random variable or a random process. It is used in information theory [84] as a measure of information-carrying capacity. Given a random variable a in a sample space X, the entropy of a is defined as

    $$I(a) = \sum_{a \in X} (prob(a) * \log \frac{1}{prob(a)}),$$

    where $prob(a)$ is the probability of a.

7.1. SYNTHESIS FOR BUILT-IN SELF TEST

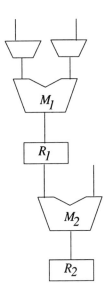

Figure 7.3: R_1 can remain a normal register if M_1s output has good randomness and M_2 has good transparency.

Intuitively, if $prob(a)$ is 0.5, then a has the maximum entropy value, carrying the most information, and is considered the most random. Therefore, each modules output signal can be associated with an entropy function to measure the randomness. The entropy value is determined by the signal probability obtained by Monte Carlo simulation.

– *transparency*: a probability approach to decide if a module can propagate any signal change at its inputs to its outputs. The transparency of a module is also obtained by Monte Carlo simulation.

Based on these metrics, if pseudorandom testability provided by a particular BILBO in the initial testable design can be replaced by normal system function, the testability overhead is reduced. This type of BIST design is called *select BIST* strategy, with sequential depth larger than 1.

For example, in the partial data path shown in Figure 7.3, if the randomness measure at register R_1 is high by calculating the output entropy of module M_1, then R_1 does not need RTPG function. Moreover, if module M_2 also has good transparency, then the error effect at R_1 can be observed at R_2 as well. So R_1 can remain a normal register without the need of MISR function, reducing BIST overhead. Since the optimized design now has sequential depth larger than 1, logic-level fault simulation is performed by SYNTEST to validate if the fault coverage is still satisfactory.

The general organization of SYNTEST environment is depicted in Figure 7.4. SYNTEST takes a VHDL behavior description as input and generates the corresponding DFG for its data path. This DFG is then scheduled by *Move Frame Scheduling* (MFS) algorithm [96], which transforms the scheduling space into a dynamic system space where moves towards the equilibrium point, guided by *Liapunov* stability theorem [11], correspond to moves in the scheduling space for deriving a schedule with balanced utilization. MFS is able to efficiently perform guided search in the huge scheduling space for design space exploration.

The testability allocator in SYNTEST then performs the above three major allocation steps in a stepwise refinement fashion to derive a testable RTL data path design. The results of several synthesized examples show that the maximal BIST strategy incurs about 20% to 27% area overhead over non-BIST design, while the select BIST strategy incurs about 2% to 6% overhead with similar fault coverage.

Post-synthesis on the RTL design by SYNTEST can follow to further trade off test area overhead, test time, and fault coverage [55].

7.1. SYNTHESIS FOR BUILT-IN SELF TEST

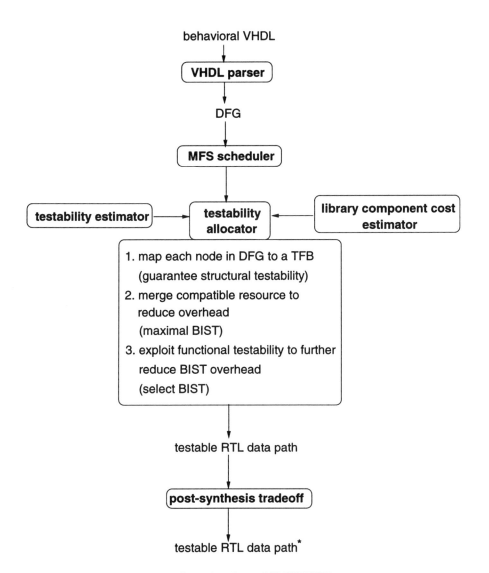

Figure 7.4: Organization of SYNTEST system.

7.1.2 RALLOC: Stanford University

RALLOC [9], a register allocation technique for BIST testability, is developed by Avra and McCluskey at Stanford University. The goal is to minimize the number of self-adjacent registers by imposing a simple constraint in the phase of register allocation, thereby reducing the area overhead of BIST circuitry.

Figure 7.5 illustrates the flow of RALLOC system. Assuming scheduling and module allocation are done, a conflict graph is first built, where each node represents a variable and each arc denotes a certain type of constraint, such as self-adjacency, with a weight as the cost. A graph coloring heuristic guided by the cost is then applied to find a small set of registers where self-adjacent registers are minimized. Notice that coloring a conflict graph is the dual problem of clique-partitioning a compatibility graph discussed in Sections 2.2.2.5 and 2.4.2.2. The self-adjacent registers that cannot be avoided by RALLOC are mapped to CBILBO registers [133], while the remaining non-self-adjacent ones are mapped to BILBO registers.

The following three major steps are performed by RALLOC to reduce self-adjacent registers during register allocation:

- step 1: Construct a *register conflict graph* from the input SDFG description to represent testability constraint.

  In a register conflict graph, a node represents a variable in the SDFG, and a *conflict edge* between two nodes indicates that these two corresponding variables cannot be assigned to the same register.

  Figure 7.6(b) gives an example of a register conflict graph for the SDFG shown in Figure 7.6(a). In the register conflict graph, nodes a and b, for example, have a solid-line edge to indicate the conflict due to the overlapping lifetimes of a and b. Nodes a and c have a dashed edge because a is an input of module $+_1$ while c is the output of the same module, and assigning a and c to the

7.1. SYNTHESIS FOR BUILT-IN SELF TEST

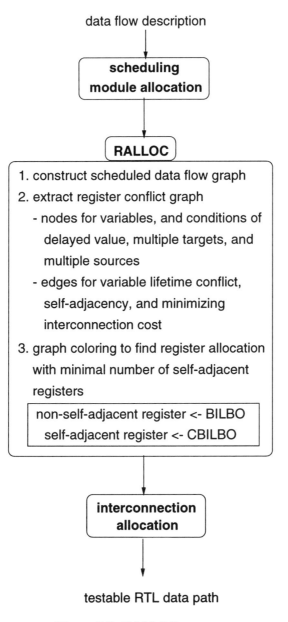

Figure 7.5: RALLOC system.

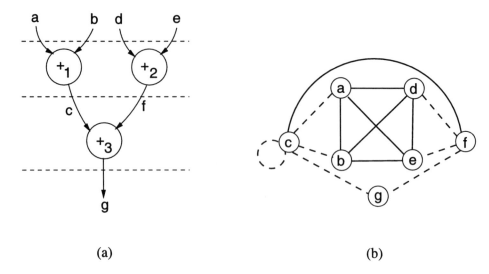

Figure 7.6: (a) An SDFG example; (b) its register conflict graph.

same register results in self-adjacency. This kind of dashed edge is called *self-adjacent constraint edge*. Furthermore, a self-edge is associated with node c if modules $+_1$ and $+_3$ are assigned to the same adder. In this case, the register allocated to c becomes both the input and the output of the adder, resulting in self-adjacency that cannot be avoided by register allocation. Therefore, a CBILBO register must be allocated to c.

- step 2: Add cost-constraint edges to the register conflict graph.

  Some extensions to the above register conflict graph can be considered to further facilitate register allocation for testability and to minimize interconnection during register allocation:

  – In order to increase the possibility of assigning variables to non-self-adjacent registers, multiple nodes are created in the register conflict graph for a single variable in the SDFG under three conditions: *delayed values, multiple targets,* and *multiple sources,* as illustrated in Figure 7.7.

7.1. SYNTHESIS FOR BUILT-IN SELF TEST

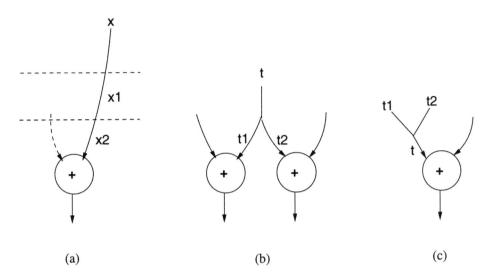

Figure 7.7: (a) delayed values; (b) multiple targets; (c) multiple sources.

* *Delayed values* refer to the situation where a variable has a separate node in the register conflict graph for each cycle of its lifetime. In this case, the variable can be stored in different registers in different cycles of its lifetime if it is beneficial. This is similar to the idea of the extended binding model presented in [68].

* *Multiple targets* means that a node is added to the register conflict graph for each fanout branch of a variable across a clock boundary in the SDFG.

* *Multiple sources*, which exists as a result of a conditional statement, means that a node is added to the register conflict graph for each input of a module that represents the same variable.

After these nodes are added, necessary conflict edges are added to the register conflict graph accordingly to reflect the correct allocation constraints.

- In order to minimize interconnection cost, a weighted *cost edge* is added between two nodes that do not have conflict edges to guide the graph coloring algorithm performed next. For example, for the SDFG of Figure 7.6, if modules $+_1$ and $+_3$ are assigned to the same adder, variables a and f should have a cost edge with negative weight in the register conflict graph to indicate the high preference of sharing the same register to reduce interconnection cost.

- step 3: Apply graph coloring algorithm to perform register allocation

    A modified version of *Brelaz* graph coloring algorithm [128] was implemented to color the register conflict graph for register allocation. Figure 7.8 outlines the graph coloring algorithm used by RALLOC [10]. The algorithm sorts the uncolored nodes in the register conflict graph according to the four priority criteria listed in Figure 7.8. Then the first uncolored node in the sorted list is selected to be assigned the color of an adjacent node with the lowest negative-weight cost edge.

To evaluate the resulting RTL data path, the following cost function is used to measure:

$$cost_{BIST} = 20\text{nsa-reg} + 35\text{sa-reg} + \#\text{mux-in} + \#int + \#ctl,$$

where #nsa-reg is the number of non-self-adjacent registers to be implemented by BILBOs of size 20 units, #sa-reg is the number of self-adjacent registers to be implemented by CBILBO of size 35 units, #mux-in is the number of multiplexer inputs, $\#int$ is the number of interconnects, and $\#ctl$ is the number of control signals. Experimental results for several examples show that the data path derived without consideration of minimizing self-adjacent registers can have higher $cost_{BIST}$ than that derived by RALLOC.

```
graph_coloring(G: register conflict graph)
{
    while (there are uncolored nodes in G) {
        sort uncolored nodes based on the following priority:
            1. self-adjacent conflict edge
            2. least number of available colors
            3. most number of conflict edges
            4. least weight sum of cost edges
        n = node with the highest priority;
        if (n has adjacent node via negative-weight edge) {
            /* have sharing preference */
            n' = adjacent node of n via the lowest negative-weight edge;
            assign n the color of n';
        } else
            assign n the color of a non-adjacent node;
    }
}
```

Figure 7.8: Graph coloring algorithm for register allocation in RALLOC.

7.1.3 SYNCBIST: University of California, San Diego

SYNCBIST (SYNthesis for Concurrent BIST) [57, 58] is developed by Harris and Orailoğlu at University of California, San Diego. The objective is to synthesize a BIST testable register-transfer data path with high test concurrency. The resulting data path contains *test registers* to perform RTPG, MISR, or both functions, and a BIST test plan for testing the design. *Partial-intrusion* BIST strategy [2] is assumed, where only a subset of the registers are used for RTPG or MISR functions. This is similar to the *select BIST* strategy used in SYNTEST discussed in Section 7.1.1, with the motivation of reducing circuit complexity and delay.

To exploit test concurrency, the *test paths* in a circuit are identified and scheduled appropriately so that every module is tested and the total test time is minimized. A test path is a sub-circuit of a data path through which test data propagate. If two or more test paths share some hardware, these test paths are said to *conflict* and cannot be tested concurrently, thereby increasing test time. Such test conflict caused by hardware sharing must be avoided during high-level synthesis to improve test concurrency. Test conflicts in a data path can be further classified into *hard conflict* and *soft conflict*:

- Hard conflict occurs when a register in one test path is used as an RTPG while in another test path the same register is used as a MISR. Figure 7.9(a) gives such an example of hard conflict.

    In this case, a CBILBO implementation could be used to resolve the hard conflict. But because using CBILBO registers imposes large area/delay overhead, SYNCBIST does not take this approach. Instead, each test path involved in the hard conflict is assigned a different test session. So the maximal number of hard-conflicting test paths gives the lower bound on the number of test sessions.

7.1. SYNTHESIS FOR BUILT-IN SELF TEST

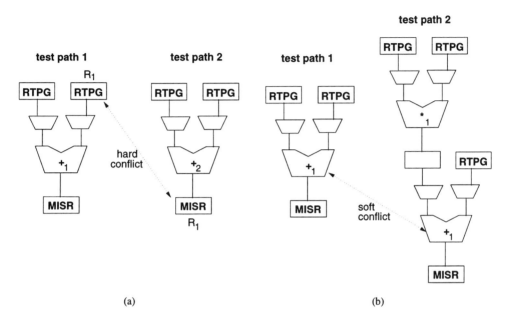

Figure 7.9: (a) Hard conflict: the same registers are used for both RTPG and MISR; (b) soft conflict: $+_1$ is shared in both test paths.

- Soft conflict occurs when different test paths share intermediate registers (not used as an RTPG or MISR), modules, multiplexers, or interconnects at the same cycle. This can be avoided by scheduling conflicted resources into different cycles.

For example, Figure 7.9(b) has the adder $+_1$ shared by both test path 1 and test path 2, causing a soft conflict. This can be resolved if $+_1$ in test path 1 is scheduled in cycle 1 while $+_1$ in test path 2 is scheduled in cycle 2. However, the number of test cycles in a test session can be increased.

During high-level synthesis, an estimate of test conflicts is used to evaluate the testability of proposed scheduling and allocation decisions, and thus guide the algorithm towards designs with superior testability characteristics. SYNCBIST attempts to maximize test concurrency by avoiding test conflicts due to hardware sharing during high-level synthesis. Two metrics for estimating a modules testability are proposed, based on analysis of the reachability of each modules port from the IO pins and the test registers:

- *conflict probability*: the probability that at least two input registers of the module are included in different test paths which causes test conflicts.

- *coverage probability*: the probability that at least one input register of the module is included in a test path which connects to a test register.

To evaluate the testability quality of a partially synthesized data path during high-level synthesis, SYNCBIST examines the ratio between the average coverage probability and the maximal conflict probability over all the IO ports of a module.

SYNCBIST decomposes the test synthesis problem into the following four components and solves them sequentially:

- step 1: scheduling/allocation. Given a DFG as input, SYNCBIST performs scheduling and then allocation under performance and area constraints. The testability metrics discussed above are used to minimize the number of test conflicts in the final design. Furthermore, the scheduling and allocation algorithms perform greedy one-step look-ahead in each iteration.

    - during scheduling, SYNCBIST selectively puts clock boundaries between nodes of a DFG if the test path options for these nodes are limited.

    This causes the insertion of registers between the modules during the next allocation phase. Since these modules are adjacent to registers, they can potentially be converted to RTPGs or MISRs during structural synthesis in step 2.

7.1. SYNTHESIS FOR BUILT-IN SELF TEST

– during allocation, interconnections among modules are distributed to increase the test path options for each module without causing conflicts between the paths.

- step 2: test register selection. Structural synthesis is then applied to select a set of test registers to perform RTPG or MISR functions. The goal is to maximize test concurrency while satisfying area constraint.

  Initially, all registers are converted to test registers. Then an incremental pruning process is performed for each register to remove away such test register conversion if test concurrency is maintained. Otherwise, the register is mapped to an RTPG or MISR to improve test concurrency, depending on whether a controllability point or observability point is needed.

- step 3: test path definition. After the testable design is synthesized, the test paths need to be defined for propagating test data through the data path. On each test path, the input of a module can receive the data only from a single source at a time. Propagation through a specific test path is ensured by controlling the select signals of multiplexers using a test controller.

- step 4: test scheduling. Test schedule of each path needs to be determined to ensure two different paths do not share hardware at the same clock cycle. A probability model is devised to guide test scheduling.

Experimental results in [57] show that SYNCBIST can achieve a single test session for several examples, independent of the synthesis constraints given or the existence of operation chaining.

7.2 Synthesis for Scan Path and Test Point Insertion

Scan path is an effective test strategy which provides direct controllability and observability of memory elements by converting the memory elements into scan registers and connecting them in a scan chain as a long shift register to scan in test vectors and scan out test responses. To reduce the area and delay overhead imposed by the scan registers, partial scan techniques to select a small subset of registers for scan have been proposed mainly based on sequential loop breaking, as discussed for PHITS-PS in Chapter 4. On the other hand, the technique of test point insertion can also effectively enhance controllability and observability of internal signals using extra logic gates instead of scan registers. In this section, three high-level test synthesis systems based on partial scan and test point insertion strategies are reviewed: Genesis in Section 7.2.1, BETS in Section 7.2.2, and TBINET in Section 7.2.3.

Genesis focuses on synthesis for hierarchical testability, which exploits during high-level synthesis the data transfer paths in a design that are sensitizable for test data justification and propagation. The concept of using a sensitizable path for transferring test data is similar to the technique of finding the *identity path* (I-path) [1] or the *fault path* (F-path) [44] in a data path during test pattern generation. If such a data transfer path from primary inputs to primary outputs can be identified during synthesis for a module under test, a system-level test set can be assembled by combining the module test set, which is precomputed, and the justification/propagation sequence defined by the data transfer path. If such a path cannot be found purely by high-level synthesis, test points of multiplexers are inserted to connect to primary inputs or primary outputs to facilitate testing of the

7.2. SYNTHESIS FOR SCAN PATH AND TEST POINT INSERTION

module. Note that since word-level information is used, system-level test assembly is fast and independent of data path bit-width when compared with gate-level ATPG, which suffers seriously from large data path bit-width.

BETS performs scheduling and allocation simultaneously for a design to minimize the cost of area, delay, and partial scan using the concept of sequential loop breaking. TBINET also aims at partial scan during allocation, using heuristics based on network flow and ILP formulations, to minimize the formation of sequential loops and maximize the allocation of IO registers, which is similar to the concept discussed for PHITS-NS in Chapter 3.

7.2.1 Genesis: Princeton University

Bhatia and Jha at Princeton University developed a high-level test synthesis system called Genesis [19, 20], aiming for hierarchical testability. The purpose of *hierarchical testing* is to take advantage of the hierarchical representation of a design to speed up justification and error propagation procedures [1, 7, 26, 69, 74, 94]. The test set of each module is precomputed at gate level, assuming the inputs and outputs of the module are primary inputs and primary outputs, respectively. Then the RTL information of the circuit surrounding the module under test, which is readily available during high-level synthesis, is used to justify the precomputed test set for each module at the circuit primary inputs and to propagate the error to the circuit primary outputs. So hierarchical testing uses the hierarchical design information to speed up test generation when compared with conventional test generation which works at gate level only.

For example, given the netlist of an adder, if the high-level functionality of addition is not available, error propagation is performed at gate level, which can be very time-consuming especially when the adder has wide bit-width. However, if test generation knows the netlist is an adder, it can take advantage of such high-level

170   CHAPTER 7. STATE-OF-THE-ART HIGH-LEVEL TEST SYNTHESIS

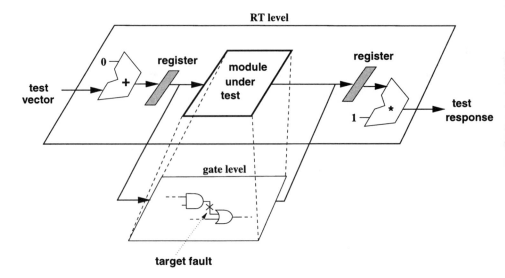

Figure 7.10: The precomputed test vector for the gate-level target fault is justified at the primary inputs by setting 0 to one operand of the adder; the test response is propagated to the primary outputs by setting 1 to one operand of the multiplier.

information to propagate the error from one operand to the output by assigning 0, the identity value of addition, to the other operand. Therefore gate-level complexity of the adder does not impact error propagation.

Figure 7.10 illustrates the concept of such a hierarchical testing approach with the assumption that the gate-level test set for the module under test is precomputed. By using RTL circuit information, a test vector can be justified at the primary inputs by setting one of the adders operands to 0, while the test response can be propagated to the primary outputs by setting one of the multipliers operands to 1.

Despite the advantage of using hierarchical design information, hierarchical testing still suffers from searching the enormous solution space to justify and propagate values. Genesis was therefore proposed to tackle this problem by confining the search space during high-level synthesis. Given a gate-level test set for each module, Genesis guarantees that a sensitizable path through a module embedded in the

7.2. SYNTHESIS FOR SCAN PATH AND TEST POINT INSERTION

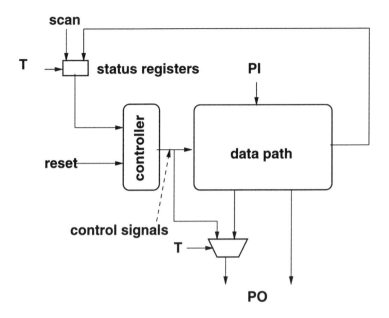

Figure 7.11: Underlying data path/controller architecture assumed by Genesis.

data path is established during synthesis such that the module can receive the test set from the circuits primary inputs and the module outputs can be observed at the circuits primary outputs. The complete system-level test set is therefore obtained as a byproduct of high-level synthesis in much shorter time when compared with gate-level sequential ATPG for the same circuit. Furthermore, such system-level test generation is not impacted by the increase of data path bit-width.

Genesis first applies a hierarchical testability analysis to identify symbolically sensitizable paths for each element in the data path described in an SDFG. Then an allocation algorithm integrated with such hierarchical testability analysis is performed to synthesize the data path and the controller with a complete system-level test set. For the element under consideration, if no such sensitizable path can be found purely by the allocation technique, a multiplexer is added to enhance controllability/overservability.

172   CHAPTER 7. STATE-OF-THE-ART HIGH-LEVEL TEST SYNTHESIS

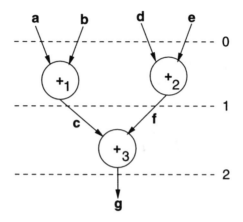

Figure 7.12: An SDFG example.

Genesis does not assume any scan of the control signals. So the original control flow defined by the SDFG is followed for value justification and propagation. However, it does assume the controller can be reset to its initial state at any time during test application. Figure 7.11 shows the underlying data path/controller architecture [18] assumed by Genesis, which allows test generation for the data path and the controller to be considered separately. For the data path, the test sequence is assembled according to the original control flow during allocation. To test the controller, gate-level sequential ATPG is applied with the assumption that the status registers are scannable, and the outputs of the controller (or the control signals) are observed at the primary outputs through a multiplexer with the test mode signal T set appropriately.

Given an SDFG, Genesis attempts to identify a *test environment*, similar to the concept of I-path [1] and F-path [44], for at least one operation bound to a module in the data path. A test environment for an operation is a set of data transfer paths that can justify the precomputed test set of the module allocated to the operation at the primary inputs and propagate the modules test response to the primary outputs, according to the control flow defined in the SDFG. A module is said to be

7.2. SYNTHESIS FOR SCAN PATH AND TEST POINT INSERTION 173

testable if at least one operation bound to this module can find a test environment. The SDFG shown in Figure 7.12 is used as an example to illustrate the idea of test environment, where variables a, b, d, e are the primary inputs, and g is the primary output. To test the module allocated to $+_2$, we can control the primary inputs d and e but need to observe the intermediate variable f. In order to observe f at the primary output g, c needs to be set to 0, which requires both the primary inputs to be set to 0. So the test environment of $+_2$ includes all the data transfer paths in the SDFG.

To demonstrate the effect of allocation on hierarchical testability, the SDFG example in Figure 7.12 is used again. Assuming variables c and g have to be assigned to the same register, two legal architectures are derived after allocation, as shown in Figure 7.13. For the architecture in Figure 7.13(a), adder $+_{1,3}$ is fully tested by controlling a to 0 and b to the test vector in the first cycle and observing the response c at g in the second cycle. Similarly, for adder $+_2$, it is fully tested by controlling d to 0 and e to the test vector in the first cycle, propagating the response f to R_2 in the second cycle. f can then be transferred to g in the third cycle through $+_{1,3}$ for observation by making c 0 in the second cycle if a and b are set to 0 in the first cycle. Notice that the control flows of testing the two adders follow what is defined in the SDFG.

For the architecture in Figure 7.13(b), adder $+_1$ can be fully tested using the control flow, where in the first cycle a is set to 0 and b is set to the test vector and in the second cycle the result c is observed at g. To test adder $+_{2,3}$, d is set to 0 and e is set to the test vector in the first cycle; the result f is stored in R_4 in the second cycle, according to the control flow defined by the SDFG. But in order to further propagate f to R_3 for observation at g, in the third cycle the test response would be transferred through the faulty $+_{2,3}$ again, which, however, may corrupt the response. So full testability cannot be guaranteed. One may argue

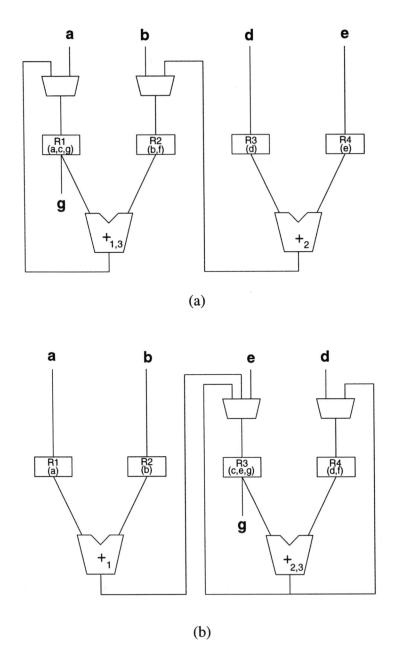

Figure 7.13: Two architectures (a) and (b) synthesized from the same SDFG in Figure 7.12 with different hierarchical testability.

7.2. SYNTHESIS FOR SCAN PATH AND TEST POINT INSERTION

that in the second cycle the test response of $+_{2,3}$ can be deviated to R_3 so that it can be observed at g. But such a deviated data transfer would require an illegal control flow not defined in the SDFG. Since Genesis does not assume scan for the control signals, the architecture in Figure 7.13(b) is not considered testable. It is the testable architecture in Figure 7.13(a) that Genesis attempts to synthesize.

The allocation algorithm of Genesis for hierarchical testability is outlined as follows:

- step 1: Build a compatibility graph for the given SDFG.

  As discussed in Section 2.4.2.2, the compatibility graph is a weighted graph where each node corresponds to a variable or an operation in the SDFG. Two nodes i and j in the compatibility graph are connected by an edge if the two corresponding variables or operations are compatible. That is, these two variables (operations) do not have overlapping lifetimes (execution times). The weight associated with an edge represents the preference of assigning the two corresponding variables (operations) to the same register (module). The weight is calculated based on the formula discussed in [20] with the main objective of area reduction.

- step 2: Perform register allocation and module allocation simultaneously by selecting an edge in the compatibility graph with the highest weight and merging the two nodes connected by the selected edge into a single compound node.

  This step is similar to the heuristic introduced in Section 2.4.2.2 to incrementally merge nodes to find clique partitioning of a compatibility graph. So the elements merged in a compound node will share the same resource.

- step 3: Evaluate hierarchical testability of the partially allocated circuit by checking if the test environment of each compound node can be identified.

If test environment cannot be found for any operation, the last allocation is undone and step 2 is repeated to select the next best edge in the compatibility graph for allocation. If the circuit is still untestable after all edges have been tried and all allocation steps are undone, extra multiplexer connections are added appropriately between primary inputs and non-input registers to enhance controllability, and/or between primary outputs and non-output registers to enhance observability.

- step 4: Update the weights of the edges in the compatibility graph.

- step 5: Repeat steps 2 through 4 until there is no edge left in the compatibility graph.

Experimental results show that Genesis can synthesize fully testable benchmark circuits with the system-level test sets generated in a matter of seconds even when the data path bit-width is 64, whereas the gate-level sequential ATPG cannot completely handle bit-widths over 8 for the same benchmarks.

7.2.2 BETS: NEC C&C Research Laboratories

The test synthesis system BETS [41] proposed by Dey, Potkonjak, and Roy explores the relationship among hardware resource sharing, sequential loops in the data path, and partial scan overhead during high-level synthesis. BETS takes a CDFG as input and breaks the loops in the CDFG with a minimal number of scan registers. Then scheduling and allocation are performed to avoid sequential loop formation in the synthesized data path by sharing the scan registers while ensuring high resource utilization.

The underlying hardware model assumed by BETS is the *dedicated register file* model, as is used by the DSP high-level synthesis systems Cathedral [129] and

7.2. SYNTHESIS FOR SCAN PATH AND TEST POINT INSERTION 177

HYPER [113]. In this model, the registers are grouped in a certain number of register files, and each register file can send data to exactly one module, while each module can send data to any register files.

Four types of loops that can be formed during synthesis are addressed by BETS:

- CDFG loop: This loop is due to a cyclic data flow imposed by a loop construct in the behavioral specification. Figure 4.4 illustrated an example of CDFG loop.

- assignment loop: This is a sequential loop caused mainly by allocation when two variables in the same data flow path share the same registers. For example, in Figure 4.1(a), since variables a and d are assigned to the same register, an assignment loop is created.

- sequential false loop: This is a sequential loop that cannot be sensitized under normal control conditions. But since the control signals to the data path are assumed fully controllable using full scan by BETS, the false loops act as real assignment loops during test generation.

- register file clique: This loop is due to the structure of the dedicated register file model.

Consider that in Figure 7.14 register R_1 in register file RF_1 and register R_2 in register file RF_2 provide operands to the adder. Based on the dedicated register file model, this adder can send the result to any register files, including RF_1 and RF_2. In this case, all registers in RF_1 and RF_2 are connected in the corresponding data path circuit graph, which forms a clique of size $m+n$ if m is the size of RF_1 and n is the size of RF_2. However, breaking all the loops in a clique of size k requires scanning $k-1$ registers, which makes partial scan very expensive.

178   CHAPTER 7. STATE-OF-THE-ART HIGH-LEVEL TEST SYNTHESIS

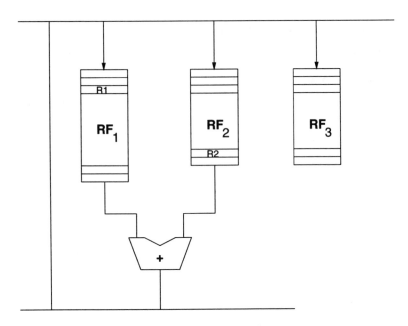

Figure 7.14: Dedicated register file model with a register file clique.

BETS starts with the CDFG of a design and the necessary resource determined by HYPER. It first selects a subset of variables to assign to scan registers for breaking all the CDFG loops. Then it simultaneously schedules and allocates modules to the operations in the CDFG using global resource utilization and testability measures. These two steps are described in more detail as follows:

- step 1: Break CDFG loops with minimal number of scan registers.

    In contrast to the PHITS approach discussed in Chapter 4 where a subset of boundary variables is selected to break CDFG loops, BETS considers all the variables in the CDFG as possible scan variable candidates to increase the opportunity of sharing scan registers.

    This step attempts to select a set of scan variables in order to:

    – break all the loops except self-loops in the CDFG;

7.2. SYNTHESIS FOR SCAN PATH AND TEST POINT INSERTION

- assign the selected scan variables to a minimal number of scan registers;
- allocate these scan registers to other variables to break the sequential loops in the data path formed by subsequent steps.

A random walk-based approach is implemented for this step with two measures [42]: loop cutting effectiveness measure and hardware sharing effectiveness measure.

After the scan variables are selected to break the CDFG loops, a minimum set of scan registers is allocated for all these scan variables. These scan registers are selected from as many register files of as many modules as possible. The goal is to increase the chances of allocating these scan registers to other variables to avoid loop formation in the next step.

- step 2: Perform simultaneous scheduling and allocation for resource utilization and testability.

As listed in Figure 7.15, the synthesis algorithm of BETS iterates until all the operators in the CDFG are assigned, and then assigns all variables to registers.

The cost of assigning an operation o in the CDFG to a particular module scheduled in a particular cycle is a function of:

- testability cost $cost_{test}$: depends on the hardware cost by using scan registers to break loops and the size of loops not broken because maximal number of scan registers allowed is reached;
- resource utilization cost $cost_{RU}$: depends on the difficulty of performing scheduling and allocation for o, the likelihood of reusing interconnection, and the likelihood of reusing registers;

```
schedule_and_assign() {
    while there exists a node in CDFG not scheduled and allocated {
        o = unscheduled operation in CDFG with the least slack;
        calculate the cost of each possible schedule and module allocation
            for o, in terms of testability ($cost_{test}$), resource utilization ($cost_{RU}$),
            and flexibility for scheduling and allocation ($cost_{flex}$):
        choose the minimal cost schedule and module allocation for o;
        assign input variables of o to scan register/register files;
        update the loop information of the data path and register files;
        update the slacks of unscheduled operators;
    }
    assign variables in the register files to registers;
}
```

Figure 7.15: Test synthesis algorithm of BETS.

- scheduling and allocation flexibility $cost_{flex}$: depends on to what extent the scheduling and allocation for o can adversely affect the flexibility for scheduling and allocation for the subsequent operations in the transitive fanout of o in the CDFG.

The experimental results in [41] on several example circuits show that BETS uses much fewer scan registers to achieve high fault efficiency when compared with conventional high-level synthesis algorithm followed by a gate-level partial scan tool OPUS [35].

In addition, Dey and Potkonjak [40, 111] proposed a *hot potato transformation* technique to facilitate test synthesis by BETS. This technique is applied before BETS to transform the CDFG by introducing **deflection operators**. A deflection operation is an operation with one operand fixed to the identity value of the operation so that the output of the operation preserves the same value of the other operand. For example, addition has 0 as its identity, so an addition with a 0 operand can be treaded as interconnection and added anywhere in a CDFG without changing

7.2. SYNTHESIS FOR SCAN PATH AND TEST POINT INSERTION

the functionality of the computation.

Using the hot potato transformation technique, the original CDFG is first transformed by adding suitable deflection operations so that the number of scan registers needed to break CDFG loops is minimized. Then a second set of deflection operations is added so that the selected scan registers can be effectively used to avoid formation of sequential loops during allocation. BETS is then applied to the transformed CDFG to synthesize a testable data path with even fewer scan registers than BETS alone is applied.

The experimental results in [40] show that by adding deflection operations, BETS can indeed reduce the number of scan registers in the data path while at the same time reducing the number of resources, such as registers and interconnects.

7.2.3 TBINET: University of Wisconsin

TBINET (Testability-constrained BInding using NETwork flows) [91], developed by Mujumdar, Jain, and Saluja at University of Wisconsin, Madison, is a module and register allocation tool that can consider both area and testability. A heuristic solution is proposed that maps allocation onto a sequence of minimum cost network flow problems. The cost function is based on the penalty of forming sequential loops and registers lacking controllability or observability during synthesis, in addition to the interconnection cost. Note that only non-scan test strategy is assumed.

Given an SDFG, TBINET performs simultaneous module and register allocation for each cycle in sequence, starting with the first. The allocation for each cycle is mapped to a minimum cost network flow problem, which will be illustrated in the following example. Suppose we want to allocate for two additions add_1 and add_2 scheduled in the same cycle, and for their output variables. The given resource constraint is two adders $+_1$, $+_2$ and two registers R_1, R_2. Figure 7.16(a) shows

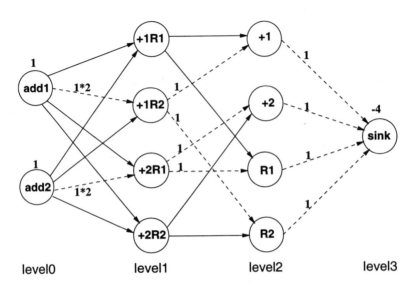

Figure 7.16: Example network with a feasible flow solution, where the arc labels allow multipliers and the dashed arcs correspond to a feasible allocation.

the complete network to find the allocation solution, where the two level-0 nodes correspond to the two additions, while the level-3 sink node is used to terminate the flow. Each addition can consider any of the four adder-register pairs for binding: $+_1R_1$, $+_1R_2$, $+_2R_1$, and $+_2R_2$. So, four level-1 nodes corresponding to the adder-register pairs are used and have arcs connected from each level-0 node. A supply of one flow unit is assigned to each level-1 node, with the constraint that each addition is assigned to only one adder-register pair. Each level-0 to level-1 arc indicates a possible binding and has a cost associated with it. The cost, to be defined later, accounts for the area of interconnection and testability, such as sequential loops. But in order to derive the complete network in Figure 7.16(a) for correct allocation, several modifications have been added:

- To constrain no more than one addition (level-0 node) to be mapped to the same adder-register pair (level-1 node), a set of level-2 nodes is introduced,

7.2. SYNTHESIS FOR SCAN PATH AND TEST POINT INSERTION

one for each adder and register, as shown in Figure 7.16(a). An arc is put with zero cost and one unit capacity from each level-1 adder-register pair node to the corresponding level-2 adder node and register node.

- Now, the outgoing flow of a mapped level-1 node requires two units, one to an adder node and one to a register node. So the incoming flow to the mapped level-1 node needs to be two units in order to conserve the flow. The solution is to have a multiplier of 2 on the incoming arc to a mapped level-1 node, as shown in Figure 7.16(a).

- To prevent a level-2 node from having the incoming flow more than one unit, a level-3 sink node is added and the the capacity of the arc from each level-2 node to it is constrained to one.

The dashed arcs in the network in Figure 7.16 correspond to one feasible allocation. That is, add_1 is bound to $+_1$ with the output bound to R_2, while add_2 is bound to $+_2$ with the output bound to R_1.

However, the network with flow multipliers is known as a generalized network, and to find its optimal integer flow solution is an NP-complete problem [46]. To develop an efficient allocation algorithm, the network is therefore simplified by assigning two units of flow supply to each level-0 node and relaxing the level-0 to level-1 flow capacity constraint to two. Because it does not have flow multiplier, a network simplex method [17] can solve the optimal integer flow problem in polynomial time.

However, the optimal integer flow solution found by the network simplex method may not be a feasible allocation. This is because now the level-0 to level-1 arc can have 0, 1, or 2 units of flow, but the feasible allocation requires the flow be either 0 or 2 units. This requirement cannot be imposed by the network simplex method.

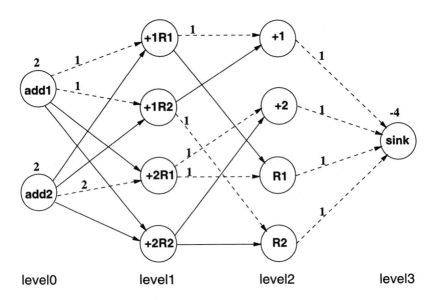

Figure 7.17: Infeasible binding solution of the simplified network, as indicated by dashed arcs.

So an optimal solution like the one in Figure 7.17 can be derived, where the binding of add_1 is not feasible.

In the case of an allocation solution with some feasible bindings and some infeasible, only the feasible bindings are preserved. The remaining infeasible bindings are then converted to 0-1 ILP formulation to find an optimal feasible solution. The ILP formulation defines:

- O: the set of unbound operations in the cycle under consideration.

- M: the set of available modules.

- R: the set of available registers.

- $x_{o,m,r} = \begin{cases} 1 & \text{if operation } o \text{ is executed by module } m \text{ and the result is} \\ & \text{stored in register } r; \\ 0 & \text{otherwise.} \end{cases}$

7.2. SYNTHESIS FOR SCAN PATH AND TEST POINT INSERTION

- $c_{o,m,r}$: the cost associated with $x_{o,m,r}$.

The allocation problem for the unbound operations in O can therefore be expressed in ILP as

$$\text{minimize} \sum_{o \in O} \sum_{m \in M} \sum_{r \in R} c_{o,m,r} \times x_{o,m,r} \quad (7.1)$$

subject to the constraints

$$\sum_{\substack{m \in M \\ o \leftrightarrow m}} \sum_{r \in R} x_{o,m,r} = 1, \ \forall o \in O \quad (7.2)$$

$$\sum_{\substack{o \in O \\ o \leftrightarrow m}} \sum_{r \in R} x_{o,m,r} \leq 1, \ \forall m \in M \quad (7.3)$$

$$\sum_{\substack{o \in O \\ o \leftrightarrow m}} \sum_{m \in M} x_{o,m,r} \leq 1, \ \forall r \in R \quad (7.4)$$

where $o \leftrightarrow m$ means operation o can be executed by module m. The objective function 7.1 wants to minimize the total binding cost. Constraint 7.2 states that each operation must be bound to exactly one operation. Constraint 7.3 states that each module can be allocated to at most one operation in the same cycle. Similarly, constraint 7.4 says each register can be allocated to at most one variable in the same cycle.

The major steps of the overall TBINET test synthesis algorithm are then outlined below:

- step 1: Build a simplified network for the operations scheduled in the current cycle.

- step 2: Calculate the cost associated with each arc in the simplified network. The arc cost is the sum of the following costs based on the partial design:

    - cost of wires and multiplexers;

- penalty cost if new connections form self-loops as well as non-self-loops (or sequential loops);
  - penalty cost if the registers are not input or output registers.

- step 3: Solve the network flow problem in polynomial time using the network simplex method.

- step 4: Update the partial design with feasible binding and interconnection derived from the network flow solution.

- step 5: If there are infeasible operation bindings,
  - use ILP formulation to solve the binding of these operations, and update the partial design with the ILP result.

The experimental results evaluated by sequential ATPG show that increasing the number of input/output registers can increase the fault coverage and reduce the test length. However, because the effect of sequential loops (or non-self-loops) is larger, reducing the number of self-loops alone does not have significant impact on testability.

7.3 Test Synthesis at RT Level

At register-transfer level although the circuits architecture is determined, the structural information available can still help test synthesis algorithms to economically add DFT circuitry with less overhead when compared with conventional gate-level DFT approaches. Below, two effective RTL test synthesis methods are discussed: ADEPT in Section 7.3.1, and the Chen-Karnik-Saab method in Section 7.3.2

7.3.1 ADEPT: University of Illinois

The ADEPT system [34] proposed by Chickermane, Lee, and Patel at University of Illinois at Urbana-Champaign performs intelligent partial scan flip-flop selection based on circuits RTL information, which is usually unavailable to gate-level DFT tools. Experimental result shows that the partial scan selection by ADEPT is more economical than gate-level result and improves both gate-level and high-level ATPG performance.

ADEPT takes as input a circuits RTL description in VHDL and transforms it into an *execution flow graph* (EFG) model to calculate high-level testability measures. Then a small set of scan flip-flops is selected based on the testability measures. The circuits handled by ADEPT are assumed to be controlled by micro-instructions.

The EFG is used to capture the information derived from the RTL description, which models the circuit behavior using a CDFG with data flows enabled by micro-instructions. An EFG has three types of nodes: register nodes for memory elements, combinational nodes for combinational modules, and fan nodes to join arcs (fan-in nodes) or fork arcs (fan-out nodes). An arc from nodes i to j represents a data transfer from i to j triggered by the micro-instruction associated with the arc as the label.

Figure 7.18 shows an example of EFG for a register-counter [34]. When the micro-instruction LDCT.load fires, a data transfer from IN to the register node is enabled. For RFCT.decr, PRCT.decr, and TWB.decr three micro-instructions, the data flows from the register through the decrementer (a combinational node) back to the register. For JZ.hold and CJS.hold, the data in the register circulates back to hold the same value.

Based on the high-level circuit information modeled in the EFG, the EFG can be used to calculate each node's testability measure, called *test sequence range* (TSR). The TSR of each node in an EFG essentially represents the range in the length of

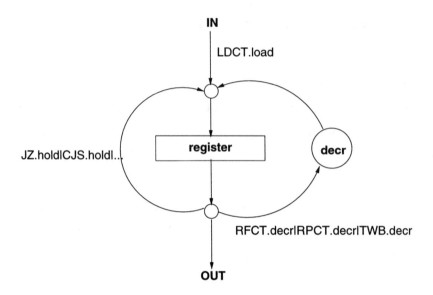

Figure 7.18: Execution flow graph for a register-counter.

the micro-instruction sequence to control and observe the node.

Several criteria are considered to evaluate the testability of an EFG for scan register selection:

- conflict source: Similar to the fanout reconvergence stem in a combinational circuit, a potential conflict source v for inputs i_1 and i_2 of a module is a node in the EFG that is in the common input cones of i_1 and i_2, and the micro-instruction sequences required to transfer data from v to i_1 and i_2 are the same. Figure 7.19 gives an example of such conflict source. In this case, one register is selected to scan by ADEPT which is on the path from v to either i_1 or i_2.

- functional constraint: Functional constraints on a signal are don't care conditions under which the signal values can never occur. For example, if a register is fixed to an 8-bit constant, all the other 255 values are invalid.

7.3. TEST SYNTHESIS AT RT LEVEL

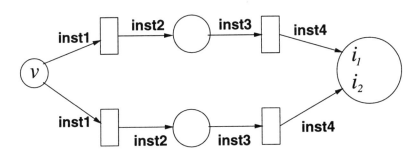

Figure 7.19: Example of conflict source.

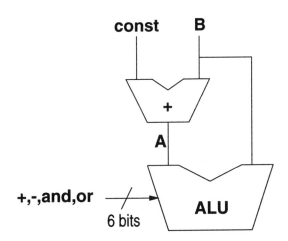

Figure 7.20: Example of functional constraint.

Other more complicated situations of functional constraints can be illustrated by the example in Figure 7.20. One of the constraints is that since only four modes of the ALU can be selected, test vectors that require the other modes are unable to be justified at the system inputs. Another constraint is that A is not a free variable with respect to B, because $A = B + const$, where $const$ is a constant. So A and B cannot be independently justified to test the ALU.

ADEPT can analyze the functional constraints and insert scan to resolve them economically.

- profit function: The profit function is defined based on the improvement of testability measure when a certain register node is selected for scan. The profit function can therefore be used to choose the scan register with the largest profit.

Experimental result shows that the scan flip-flops selected by ADEPT are fewer when compared with the result of a gate-level fault-oriented partial scan tool OPUS [35]. Improvement in fault coverage/efficiency and speedup in test generation are achieved both by a gate-level ATPG tool HITEC [95], and by an architectural-level ATPG tool ARTEST [73, 74, 75].

7.3.2 The Chen-Karnik-Saab Method: University of Illinois

Chen, Karnik, and Saab [30] developed a system for RTL test synthesis at University of Illinois at Urbana-Champaign. The system takes as input a C or VHDL RTL description of a circuit and performs three major steps to synthesize testability:

- step 1: Apply a testability analyzer, BETA [31, 32], to analyze the circuit description in the form of a CDFG.

Figure 7.21 shows a CDFG example taken from [30] for BETA to analyze testability. This CDFG has three execution paths, P_1, P_2, and P_3, which are

7.3. TEST SYNTHESIS AT RT LEVEL

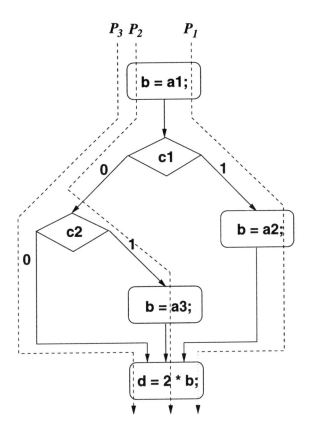

Figure 7.21: Example of a CDFG used in the Chen-Karnik-Saab method with 3 execution paths, P_1, P_2, and P_3.

first extracted. BETA then analyzes the variables on each path based on the controllability and observability. For example, on P_3, variable d is controllable if c_1, c_2, and b are controllable. If any of them is found non-controllable, d is considered non-controllable. Therefore, all the variables can be classified into groups with different testability by analyzing if their values can be easily justified at the primary inputs or propagated to the primary outputs.

- step 2: Based on BETAs analysis, iteratively select the variable with the least difficulty with value justification and propagation for test point or partial scan flip-flop insertion.

  This selection heuristic is due to the observation that when a less difficult-to-test variable is selected, some other more difficult-to-test variables tend to become testable automatically.

- step 3: Test statement insertion, an alternative to step 2, can be used to modify the circuit based on the selected test points.

  For example, suppose variable u is difficult to control but v is completely controllable. A test statement can be inserted to make u also completely controllable during test mode by assigning v to u when T is 1:

$$\texttt{if } (T == 1) \; u := v;$$

7.4 Other Work

A number of other interesting works related to high-level test synthesis are briefly overviewed in this section.

7.4. OTHER WORK

Transformation and Resynthesis for Testability

Two RTL transformations are proposed [21] by Bhattacharya *et al.* to generate canonical representation of the data path, from which exact observability don't cares for control signals can be extracted. These don't cares are used to minimize the control logic such that the network combining the data path and the control logic is fully testable, while all the false paths in the data path are also removed.

Automatic Generation of Behavioral Modification

AMBIANT developed by Vishakantaiah *et al.* [131] performs testability analysis on an RTL description based on ATKET [130] to suggest modifications targeted at improving testability of the design.

Test Synthesis Guided by Testability Measures

A testability analysis tool is presented in [43] by Fernández *et al.* to guide a high-level synthesis algorithm. The testability of a design is analyzed based on two metrics: controllability/observability and loop analysis. The controllability and observability of each signal are calculated using signal probability. The loop analysis considers both combinational and sequential false loops.

Testability Improvement for VHDL Specification

A testability analysis and improvement technique is proposed in [52] by Gu *et al.* for RTL specification in VHDL. An *extended timed petri net* (ETP) model is used to represent the design internally after the VHDL specification is compiled. Then testability analysis based on controllability and observability is performed to measure the difficulty with the ETP model for ATPG. Test improvement techniques are then applied, which employ partial scan and test point insertion strategies.

Partial scan is inserted based on breaking sequential loops, while test point insertion is based on the testability of signal lines.

Stepwise Refinement Synthesis for Easy Testability

A simultaneous scheduling and allocation algorithm for testability is proposed in [64] by Kim *et al.* without assuming any particular test strategy. Testability measure is defined based on the controllability/observability of registers, the length of sequential depth, and the number of sequential loops. An initial data path is constructed by allocating each operation a module and each variable a register. Then iterative refinement is performed to merge compatible modules and registers according to a cost function that measures testability, total execution time, and area.

Non-Scan DFT Technique

In [39], a test point insertion algorithm is proposed by Dey and Potkonjak to totally avoid using scan. An *EXU S-graph* is used to analyze sequential loops, which differs from the conventional S-graph [33] in that the node in the EXU S-graph corresponds to an execution unit, rather than a register, in the data path. Furthermore, not all the sequential loops found in the EXU S-graph are selected to break, because some of them may be testable already. A loop in the EXU S-graph is considered to break only when it has nodes whose output cannot be justified/propagated in at most $k+1$ control steps, where k is called the controllability/observability level of the loop. So, given the parameter k specified by the user, the test point is inserted iteratively at the output of an execution unit if its corresponding node in the EXU S-graph is in a loop with a controllability/observability level larger than k, and such insertion is also considered to maximally reduce the controllability/observability levels of other loops at the same time.

7.4. OTHER WORK

Design for Hierarchical Testability of RTL Circuit

Ghosh *et al.* presented in [49] a test point insertion method for hierarchical testability for an RTL design after scheduling and allocation are done. The concept of test environment proposed in Genesis in Section 7.2.1 is used to identify the bottleneck from the hierarchical controllability/observability point of view. Then a small number of multiplexers are selectively inserted to make the circuit hierarchically testable. Therefore, the benefits of hierarchical testability discussed in Section 7.2.1 hold valid with this proposed method. In the cases where more than one location are considered the candidates for multiplexer insertion, critical paths are avoided to minimize the timing impact. Since no scan is used, the system-level test set can be applied at-speed.

Macro Testing

A design style based on macro cells which reuses previous hardware investment stored in a macro cell library promises to cope with the problem with ever-increasing design complexity, delivering higher quality and productivity. Testing strategy associated with such a macro-based design methodology, as discussed in [13, 14], requires that either the test set or the means of generating such a test set exist, and that the *test protocol* be known. Test protocol describes the access to the macro from the system IO pins to test the macro as a stand-alone entity. To produce a test protocol for an embedded macro, trade-off needs to be made among possible ways to access the macro under test, such as by using the existing functional data paths or DFT circuitry like scan chains, or by adding extra logic to transfer test data.

Arithmetic BIST in High-Level Synthesis

Mukherjee *et al.* proposed in [93] an arithmetic built-in self test (ABIST) scheme for high-level synthesis of data path architecture. The ABIST scheme makes use of the arithmetic functional units, such as adders or subtractors, to perform test pattern generation [53] and test response compaction [114]. Given a DFG for high-level synthesis, random test pattern generators using arithmetic operations are first introduced. A testability measure based on random pattern coverage of the input space for each operation, called *subspace state coverage* in [93], is evaluated for resource sharing. It is preferred that two operations without conflicting execution times share the same functional unit if the random pattern coverage at the input is maximized to ensure high random testability. Then, based on measuring the observability of terminal faults of each functional unit, compactors are economically placed to collect responses from the entire DFG.

Allocation for Low BIST Area Overhead

The data path allocation method proposed by Parulkar *et al.* in [105] attempts to employ the concept of I-path [1] to reduce the area overahead imposed by BILBO registers. Since the I-path does not alter the test data transferred along it, only the registers at the head and the tail of the I-path are considered to convert to BILBO registers. If the circuit has self-adjacent registers, CBILBO registers, which have much larger area overhead, are then used to break the self-loops. The method aims at maximizing the sharing of I-paths between different functional units while at the same time avoiding the possibility of CBILBO register assignment.

7.5 Summary

In this chapter, we surveyed major recent work on high-level test synthesis other than PHITS that is based on either BIST or scan path. High-level BIST synthesis mainly attempts to reduce BILBO register overhead by minimizing the occurrence of self-adjacent registers. In addition, RALLOC considers to use CBILBO registers to implement self-adjacent registers, SYNTEST exploits the circuit functionality to minimize the required BILBO registers, and SYNCBIST takes test concurrency into account. For scan-path-based high-level test synthesis, sequential loop reduction during architecture generation is shown to be the most effective approach by BETS and TBINET. Synthesis for hierarchical testability proposed by Genesis is attractive in that the precomputed test set of each module can be used to efficiently assemble the system-level test set without suffering from the problem of large data path bit-width associated with conventional gate-level ATPG.

As the chip complexity increases and the time-to-market decreases, increasingly higher level of design abstraction becomes essential to managing the design complexity and making the design portable to reuse. Design automation from such a high level of abstraction will be facilitated by high-level synthesis. Furthermore, in order to ensure the correctness of a complex design without missing the small market window, the designer will be more willing to add testing features during the design process. Therefore, high-level test synthesis, which considers testability during high-level synthesis, is a practical solution to satisfying all these important constraints while minimizing the overhead.

Bibliography

[1] M. S. Abadir and M. A. Breuer. "A knowledge-based system for designing testable VLSI chips". In *Proc. IEEE Design & Test*, pages 56–68, Aug. 1985.

[2] M. S. Abadir and M. A. Breuer. "Test schedules for VLSI circuits having built-in test hardware". In *IEEE Trans. Comput.*, pages 361–367, Apr. 1986.

[3] M. Abramovici, M. A. Breuer, and A. D. Friedman. *Digital Systems Testing and Testable Design*. Computer Science Press, New York, 1990.

[4] M. Abramovici, M. A. Breuer, and A. D. Friedman. *Digital Systems Testing and Testable Design*, chapter 9. Computer Science Press, New York, 1990.

[5] M. Abramovici, J. J. Kulikowski, and R. K. Roy. "The best flip-flops to scan". In *Proc. Int. Test Conf.*, pages pp. 166–173, 1991.

[6] A. V. Aho, R. Sethi, and J. D. Ullman. *Compilers: Principles, Techniques, and Tools*. Addison-Wesley Company, Reading, MA, 2nd edition, 1986.

[7] P. N. Anirudhan and P. R. Menon. "Symbolic test generation for hierarchically modeled digital systems". In *Proc. Int. Test Conf.*, pages 461–469, Oct. 1989.

[8] P. Ashar and S. Malik. "Implicit computation of minimum-cost feedback-vertex sets for partial scan and other applications". In *Proc. 31st Design Automation Conf.*, pages 77–80, San Diego, June 1994.

[9] L. Avra. "Allocation and assignment in high-level synthesis for self-testable data paths". In *Proc. Int. Test Conf.*, pages 463–472, Nashville, Aug. 1991.

[10] L. Avra. *Synthesis Techniques for Built-In Self-Testable Designs*. PhD thesis, Dept. of Electrical Engineering, Stanford University, June 1994.

[11] S. P. Banks. *Control System Engineering: Modeling and Simulation*. Prentice Hall, Englewood Cliffs, NJ, 1986.

[12] P. H. Bardell, W. H. McAnney, and J. Savir. *Built-In Test for VLSI: Pseudorandom Techniques*. John Wiley & Sons, New York, 1987.

[13] F. P. M. Beenker, R. G. Bennetts, and A. P. Thijssen. *Testability Concepts for Digital ICs—The Macro Test Approach*. Kluwer Academic Publishers, Boston, 1995.

[14] F. P. M. Beenker, K. J. E. van Eerdewijk, R. B. W. Gerritsen, F. N. Peacock, and M. van der Star. "Macro testing: Unifying IC and board test". In *IEEE Design & Test of Computers*, pages 26–32, Oct. 1986.

[15] R. A. Bergamaschi and R. Camposano. "Behavioral synthesis: from research to production use". Tutorial handout in Int. Conf. Computer-Aided Design, Nov. 1994.

[16] R. A. Bergamaschi and A. Kuehlmann. "A system for production use of high-level synthesis". In *IEEE Trans. VLSI Systems*, pages 233–243, Sept. 1993.

[17] D. Bertsekas. *Linear Network Optimization*. MIT Press, Cambridge, MA, 1991.

[18] S. Bhatia. *Synthesis for Testability of Digital VLSI Circuits*. PhD thesis, Dept. Electrical Engineering, Princeton Univ., 1994.

[19] S. Bhatia and N. Jha. "Behavioral synthesis for hierarchical testability of controller/data path circuits with conditional branches". In *Proc. Int. Conf. Computer Design*, pages 91–96, Boston, Oct. 1994.

[20] S. Bhatia and N. Jha. "Genesis: a behavioral synthesis system for hierarchical testability". In *Proc. European Design & Test Conf.*, pages 272–276, Feb. 1994.

[21] S. Bhattacharya, F. Brglez, and S. Dey. "Transformations and resynthesis for testability of RT-level control-data path specifications". In *IEEE Trans. VLSI Systems*, volume 1, pages 304–318, Sept. 1993.

[22] J. Biesenack, M. Koster, T. Langmaier, S. Ledeux, S. März, M. Payer, M. Pilsl, S. Rumler, H. Soukup, A. Stoll, N. Wehn, and P. Duzy. "The Siemens high-level synthesis system CALLAS". In *IEEE Trans. VLSI Systems*, pages 244–253, Sept. 1993.

[23] R. K. Brayton, R. Camposano, G. De Micheli, R. Otten, and J. van Eijndhoven. "The Yorktown Silicon Compiler System". In D. D. Gajski, editor, *Silicon Compilation*, pages 204–310. Addision-Wesley, Reading, MA, 1988.

[24] R. K. Brayton, G. D. Hachtel, C. T. McMullen, and A. L. Sangiovanni-Vincentelli. *Logic Minimization Algorithms for VLSI Synthesis*. Kluwer Academic Publishers, Boston, 1984.

[25] R. K. Brayton, G. D. Hachtel, and A. L. Sangiovanni-Vincentelli. "Multilevel logic synthesis". In *Proc. IEEE*, volume 78, pages 264–300, Feb. 1990.

[26] M. Breuer and A. D. Friedman. "Functional level primitives in test generation". In *Proc. Trans. Comput.*, pages 223–235, Mar. 1980.

[27] R. Camposano. "Path-based scheduling for synthesis". In *IEEE Trans. CAD*, volume 10, pages 85–93, Jan. 1991.

[28] R. Camposano and W. H. Wolf, editors. *High-Level VLSI Synthesis*. Kluwer Academic Publishers, Boston, 1991.

[29] S. T. Chakradhar, A. Balakrishnan, and V. D. Agrawal. "An exact algorithm for selecting partial scan flip-flops". In *Proc. 31st Design Automation Conf.*, pages 81–86, San Diego, June 1994.

[30] C.-H. Chen, T. Karnik, and D. G. Saab. "Structural and behavioral synthesis for testability techniques". In *IEEE Trans. CAD*, volume 13, pages 777–785, Jun. 1994.

[31] C.-H. Chen, C. Wu, and D. G. Saab. "Accessibility analysis on data flow graph: An approach to design for testability". In *Proc. Int. Conf. Computer Design*, pages 463–466, 1991.

[32] C.-H. Chen, C. Wu, and D. G. Saab. "Beta: Behavioral testability analysis". In *Proc. Int. Conf. Computer Aided Design*, pages 202–205, Nov. 1991.

[33] K.-T. Cheng and V. D. Agrawal. "A partial scan method for sequential circuits with feedback". In *IEEE Trans. Comput.*, volume 39, pages 544–548, Apr. 1990.

[34] V. Chickermane, J. Lee, and J. H. Patel. "Addressing design for testability at the architectural level". In *IEEE Trans CAD*, pages 920–934, July 1994.

[35] V. Chickermane and J. H. Patel. "A fault oriented partial scan design approach". In *Proc. Int. Conf. Computer-Aided Design*, pages 400–403, Santa Clara, CA, Nov. 1991.

[36] V. Chickermane and J. H. Patel. "An optimization based approach to the partial scan design problem". In *Proc. Int. Test Conf.*, pages 377–386, Sep. 1991.

[37] S. Chiu and C. A. Papachristou. "A built-in self-testing approach for minimizing hardware overhead". In *Proc. Int. Conf. Computer Design*, pages 282–285, Oct. 1991.

[38] T. H. Cormen, C. E. Leiserson, and R. L. Rivest. *Introduction to Algorithms*. The MIT Press, Cambridge, MA, 1990.

[39] S. Dey and M. Potkonjak. "Non-scan design-for-testability of RT-level data paths". In *Proc. Int. Conf. Computer-Aided Design*, pages 184–193, Washington, D.C., Oct. 1994.

[40] S. Dey and M. Potkonjak. "Transforming behavioral specifications to facilitate synthesis of testable designs". In *Proc. Int. Test Conf.*, pages 184–193, Washington, D.C., Oct. 1994.

[41] S. Dey, M. Potkonjak, and R. K. Roy. "Exploiting hardware sharing in high-level synthesis for partial scan optimization". In *Proc. Int. Conf. Computer-Aided Design*, pages 20–25, Santa Clara, CA, Nov. 1993.

[42] S. Dey, M. Potkonjak, and R. K. Roy. "Exploiting hardware sharing in high-level synthesis for partial scan optimization". Technical report, C&C Research Labs, NEC USA, Apr. 1993.

[43] V. Fernández, P. Sánchez, and E. Villar. "High-level synthesis guided by testability measures". In *Presentation in 1st Int. Test Synthesis Workshop*, Santa Barbara, CA, May 1994.

[44] S. Freeman. "Test generation for data-path logic: the F-path method". In *IEEE J. Solid-State Circuits*, pages 421–427, Apr. 1988.

[45] D. D. Gajski, N. D. Dutt, A. C.-H. Wu, and Steve Y.-L. Lin. *High-Level Synthesis: Introduction to Chip and System Design*. Kluwer Academic Publishers, Boston, 1992.

[46] M. R. Garey and D. S. Johnson. *Computers and Intractability – A Guide to the Theory of NP-Completeness*. Freeman, New York, 1979.

[47] C. H. Gebotys and M. I. Elmasry. "Integration of algorithmic VLSI synthesis with testability incorporation". In *IEEE J. Solid-State Circuits*, volume 24, pages 409–416, Apr. 1989.

[48] A. Ghosh, S. Devadas, and A. R. Newton. "Test generation for highly sequential circuits". In *Proc. Int. Conf. Computer-Aided Design*, pages 362–365, Santa Clara, CA, Nov. 1989.

[49] I. Ghosh, A. Raghunathan, and N. K. Jha. "Design for hierarchical testability of RTL circuits obtained by behavioral synthesis". In *Proc. Int. Conf. Computer Design*, pages 173–179, Austin, TX, Oct. 1995.

[50] E. F. Girczyc, R. J. Buhr, and J. P. Knight. "Application of a subset of ADA as an algorithmic hardware description language for graph-based hardware compilation". In *IEEE Trans. CAD*, volume 4, pages 134–142, Apr. 1985.

[51] M. C. Golumbic. *Algorithmic Graph Theory and Perfect Graphs*. Academic Press, San Diego, 1980.

[52] X. Gu, K. Kuchcinski, and Z. Peng. "Testability analysis and improvement from VHDL behavioral specifications". In *Proc. European Design Automation Conf.*, Grenoble, France, Sept. 1994.

[53] S. Gupta, J. Rajski, and J. Tyszer. "Test pattern generation based on arithmetic operations". In *Proc. Int. Conf. Computer-Aided Design*, pages 117–124, San Jose, CA, Nov. 1994.

[54] L. Hafer and A. C. Parker. "A formal method for the specification analysis and design of register-transfer level digital logic". In *IEEE Trans. CAD*, volume 2, pages 4–18, Jan. 1983.

[55] H. Harmanani and C. A. Papachristou. "An improved method for RTL synthesis with testability tradeoffs". In *Proc. Int. Conf. Computer-Aided Design*, pages 30–35, Santa Clara, CA, Nov. 1993.

[56] H. Harmanani, C. A. Papachristou, S. Chiu, and M. Nourani. "SYNTEST: An environment for system-level design for test". In *Proc. European Design Automation Conf.*, pages 402–407, 1992.

[57] I. G. Harris and A. Orailoğlu. "Microarchitectural synthesis of VLSI designs with high test concurrency". In *Proc. Design Automation Conf.*, pages 206–211, San Diego, June 1994.

[58] I. G. Harris and A. Orailoğlu. "SYNCBIST: synthesis for concurrent built-in self-testability". In *Proc. Int. Conf. Computer Design*, pages 101–104, Cambridge, MA, Oct. 1994.

[59] A. Hashimoto and J. Stevens. "Wire routing by optimizing channel assignment within large apertures". In *Proc. Design Automation Workshop*, pages 155–169, 1971.

[60] C.-T. Huang, Y.-C. Hsu, and Y.-L. Lin. "Optimum and heuristic data path scheduling under resource constraints". In *Proc. Design Automation Conf.*, pages 65–70, Orlando, June 1990.

[61] C.-Y. Huang, Y.-S. Chen, Y.-L. Lin, and Y.-C. Hsu. "Data path allocation based on bipartite weighted matching". In *Proc. Design Automation Conf.*, pages 499–504, Orlando, June 1990.

[62] R. V. Hudli and S. C. Seth. "Testability analysis of synchronous sequential circuits based on structure data". In *Proc. Int. Test Conf.*, pages 364–372, Aug. 1989.

[63] K. S. Hwang, A. E. Casavant, M. Dragomirecky, and M. A. d'Abreu. "Constrained conditional resource sharing in pipeline synthesis". In *Proc. Int. Conf. Computer-Aided Design*, pages 52–55, Santa Clara, CA, Nov. 1988.

[64] T. Kim, K.-S. Chung, and C. L. Liu. "A stepwise refinement datapath synthesis procedure for easy testability". In *Proc. European Design and Test Conf.*, pages 586–590, 1994.

[65] T. Kim, J. W. S. Liu, and C. L. Liu. "A scheduling algorithm for conditional resource sharing". In *Proc. Int. Conf. Computer Aided Design*, pages 84–87, Santa Clara, CA, Nov. 1991.

[66] D. Knapp, T. Ly, D. MacMillen, and R. Miller. "Behavioral synthesis methodology for HDL-based specification and validation". In *Proc. Design Automation Conf.*, pages 286–291, San Francisco, June 1995.

[67] B. Konemann, J. Mucha, and G. Zwiehoff. "Built-in logic block observation techniques". In *Proc. Int. Test Conf.*, pages 37–41, Cherry Hill, NJ, Oct. 1979.

[68] G. Krishnamoorthy and J. A. Nester. "Data path allocation using an extended binding model". In *Proc. Design Automation Conf.*, pages 279–284, Anaheim, CA, June 1992.

[69] R. P. Kunda, P. Narain, J. A. Abraham, and B. D. Rathi. "Speed up of test generation using high-level primitives". In *Proc. Design Automation Conf.*, pages 594–599, June 1990.

[70] S.-Y. Kung, H. J. Whitehouse, and T. Kailath. *VLSI and Modern Signal Processing*. Prentice Hall, Englewood Cliffs, NJ, 1985.

[71] F. J. Kurdahi and A. C. Parker. "REAL: A program for register allocation". In *Proc. Design Automation Conf.*, pages 210–215, Miami Beach, June 1987.

[72] D. H. Lee and S. M. Reddy. "On determining scan flip-flops in partial-scan designs". In *Proc. Int. Conf. Computer-Aided Design*, pages 322–325, Santa Clara, CA, Nov. 1990.

[73] J. Lee and J. H. Patel. "A signal-driven discrete relaxation technique for architectural level test generation". In *Proc. Int. Conf. Computer-Aided Design*, pages 458–461, Nov. 1991.

[74] J. Lee and J. H. Patel. "An architectural level test generator for a hierarchical design environment". In *Proc. Int. Symp. Fault-Tolerant Computing*, pages 44–51, Montreal, June 1991.

[75] J. Lee and J. H. Patel. "An instruction sequence assembling methodology for testing microprocessors". In *Proc. Int. Test Conf.*, pages 49–58, Oct. 1992.

[76] J.-H. Lee, Y.-C. Hsu, and Y.-L. Lin. "A new integer linear programming formulation for the scheduling problem in data path synthesis". In *Proc. Int. Conf. Computer-Aided Design*, pages 20–23, Santa Clara, CA, Nov. 1989.

[77] M. T.-C. Lee, Y.-C. Hsu, B. Chen, and M. Fujita. "Domain-specific high-level modeling and synthesis for ATM switch design using VHDL". In *Proc. Design Automation Conf.*, Las Vegas, June 1996.

[78] M. T.-C. Lee, N. K. Jha, and W. H. Wolf. "A conditional resource sharing method for behavioral synthesis of highly testable data paths". In *Proc. Int. Test Conf.*, pages 744–753, Baltimore, Oct. 1993.

[79] M. T.-C. Lee, N. K. Jha, and W. H. Wolf. "Behavioral synthesis of highly testable data paths under the non-scan and partial scan environments". In *Proc. Design Automation Conf.*, pages 292–297, Dallas, TX, June 1993.

[80] M. T.-C. Lee, W. H. Wolf, and N. K. Jha. "Behavioral synthesis for easy testability in data path scheduling". In *Proc. Int. Conf. Computer-Aided-Design*, pages 616–619, Santa Clara, CA, Nov. 1992.

[81] M. T.-C. Lee, W. H. Wolf, N. K. Jha, and J. M. Acken. "Behavioral synthesis for easy testability in data path allocation". In *Proc. Int. Conf. Computer Design*, pages 29–32, Cambridge, MA, Oct. 1992.

[82] T. Ly, D. Knapp, R. Miller, and D. MacMillen. "Scheduling using behavioral templates". In *Proc. Design Automation Conf.*, pages 101–106, San Francisco, June 1995.

[83] H. De Man, J. Rabaey, P. Six, and L. Claesen. "Cathedral-II: A silicon compiler for digital signal processing". In *IEEE Design & Test of Computers*, pages 13–25, Dec. 1986.

[84] M. Mansuripur. *Introduction to Information Theory.* Prentice Hall, Englewood Cliffs, NJ, 1987.

[85] T. E. Marchok, A. El-Maleh, W. Maly, and J. Rajski. "Complexity of sequential ATPG". In *Proc. European Design and Test Conf.*, pages 252–261, Paris, Mar. 1995.

[86] P. Marwedel. "A new synthesis algorithm for the MIMOLA software system". In *Proc. Design Automation Conf.*, pages 271–277, Las Vegas, June 1986.

[87] M. C. McFarland, A. C. Parker, and R. Camposano. "The high-level synthesis of digital systems". In *Proc. IEEE*, volume 78, pages 301–318, Feb. 1990.

[88] G. D. Micheli. *Synthesis and Optimization of Digital Circuits*. McGraw-Hill, Inc., 1994.

[89] G. D. Micheli, David Ku, F. Mailhot, and T. Truong. "The Olympus synthesis system". In *IEEE Design & Test of Computers*, pages 37–53, Oct. 1990.

[90] A. Miczo. *Digital Logic Testing and Simulation*. Harper and Row, New York, 1986.

[91] A. Mujumdar, R. Jain, and K. Saluja. "Behavioral synthesis of testable designs". In *Proc. Int. Symp. Fault-Tolerant Comput.*, pages 436–445, Austin, TX, June 1994.

[92] A. Mujumdar, K. Saluja, and R. Jain. "Incorporating testability considerations in high-level synthesis". In *Proc. Int. Symp. Fault-Tolerant Comput.*, pages 272–279, Boston, July 1992.

[93] N. Mukherjee, M. Kassab, J. Rajski, and J. Tyszer. "Arithmetic built-in self test for high-level synthesis". In *Proc. VLSI Test Symp.*, pages 132–139, Princeton, NJ, Apr. 1995.

[94] B. T. Murray and J. P. Hayes. "Hierarchical test generation using precomputed tests for modules". In *IEEE Trans. CAD*, pages 594–603, June 1990.

[95] T. M. Niermann and J. H. Patel. "HITEC: A test generation package for sequential circuits". In *Proc. European Design Automation Conf.*, pages 214–218, Feb. 1991.

[96] M. Nourani and C. A. Papachristou. "Move frame scheduling and mixed scheduling allocation for the automated synthesis of digital systems". In *Proc. Design Automation Conf.*, pages 99–104, Anaheim, CA, June 1992.

[97] B. M. Pangrle. "Splicer: A heuristic approach to connectivity binding". In *Proc. Design Automation Conf.*, pages 536–541, Anaheim, CA, June 1988.

[98] B. M. Pangrle and D. D. Gajski. "Slicer: A state synthesizer for intelligent silicon compilation". In *Proc. Int. Conf. Computer-Aided Design*, pages 536–541, Santa Clara, CA, Nov. 1987.

[99] C. A. Papachristou. "Rescheduling transformations for high-level synthesis". In *Proc. Int. Symp. Circuits & Systems*, pages 766–769, Portland, OG, May 1989.

[100] C. A. Papachristou, S. Chiu, and H. Harmanani. "A data path synthesis method for self-testability designs". In *Proc. Design Automation Conf.*, pages 378–384, San Francisco, June 1991.

[101] C. A. Papachristou and H. Konuk. "A linear program driven scheduling and allocation method followed by an interconnect optimization algorithm". In *Proc. Design Automation Conf.*, pages 77–83, Orlando, June 1990.

[102] C. H. Papadimitriou and K. Steiglitz. *Combinatorial Optimization – Algorithms and Complexity*. Prentice-Hall, Englewood Cliffs, NJ, 1982.

[103] N. Park and A. C. Parker. "Sehwa: A software package for synthesis of pipelines from behavioral specification". In *IEEE Trans. CAD*, volume 7, pages 356–370, Mar. 1988.

[104] A. C. Parker, J. T. Pizarro, and M. Mlinar. "MAHA: A program for datapath synthesis". In *Proc. Design Automation Conf.*, pages 461–466, Las Vegas, June 1986.

[105] I. Parulkar, S. Gupta, and M. A. Breuer. "Data path allocation for synthesizing RTL design wtih low BIST area overhead". In *Proc. Design Automation Conf.*, pages 395–401, San Francisco, June 1995.

[106] P. G. Paulin. "DSP design tool requirements for the nineties: An industrial perspective". In *Proc. 6th Int. Workshop High-Level Synthesis*, Laguna Niguel, CA, Nov. 1992.

[107] P. G. Paulin and J. P. Knight. "Force-directed scheduling for the behavioral synthesis of ASIC's". In *IEEE Trans. CAD*, volume 8, pages 661–678, June 1989.

[108] P. G. Paulin and J. P. Knight. "Scheduling and binding algorithms for high-level synthesis". In *Proc. Design Automation Conf.*, pages 25–29, Las Vegas, June 1989.

[109] P. G. Paulin, J. P. Knight, and E. F. Girczyc. "HAL: A multi-paradigm approach to automatic data path synthesis". In *Proc. Design Automation Conf.*, pages 263–270, Las Vegas, June 1986.

[110] Z. Peng. "Synthesis of VLSI systems with the CAMAD design aid". In *Proc. Design Automation Conf.*, pages 278–284, Las Vegas, June 1986.

[111] M. Potkonjak and S. Dey. "Optimizing resource utilization and testability using hot potato techniques". In *Proc. Design Automation Conf.*, pages 201–205, San Diego, June 1994.

[112] B. T. Preas and M. J. Lorenzetti, editors. *Physical Design Automation of VLSI System*. The Benjamin/Cummings Publishing Company, Menlo Park, CA, 1988.

[113] J. M. Rabaey, C. Chu, P. Hoang, and M. Potkonjak. "Fast prototyping of datapath-intensive architecture". In *IEEE Design & Test of Computers*, pages 40–51, June 1991.

[114] J. Rajski and J. Tyszer. "Test response compaction in accumulators with rotate carry adders". In *IEEE Trans. CAD*, pages 367–370, Apr. 1993.

[115] L. Rosqvist. "Test and testability of ASICs". In N. G. Einspruch and J. L. Hilbert, editors, *Application Specific Integrated Circuit (ASIC) Technology*, chapter 8. Academic Press, San Diego, 1991.

[116] E. Roza, J. Biesterbos, B. De Loore, and J. Van Meerbergen. "On the application of architectural synthesis in the design of high-volume production ICs for consumer applications". In *Proc. 6th Int. Workshop High-Level Synthesis*, pages 2–15, Laguna Niguel, CA, Nov. 1992.

[117] E. M. Sentovich, K. J. Singh, C. Moon, H. Savoj, R. K. Brayton, and A. Sangiovanni-Vincentelli. "Sequential circuit design using synthesis and optimization". In *Proc. Int. Conf. Computer Design*, pages 328–333, Cambridge, MA, Oct. 1992.

[118] N. Sherwani. *Algorithms for VLSI Physical Design Automation*. Kluwer Academic Publishers, Boston, 1995.

[119] L. Stok. "An exact polynomial-time algorithm for module allocation". In *Proc. 5th Int. Workshop High-Level Synthesis*, pages 69–76, Bühlerhöhe, Germany, Mar. 1991.

[120] D. E. Thomas and T. E. Fahrman. "Industrial uses of the system architect's workbench". In R. Camposano and W. H. Wolf, editors, *High-Level VLSI Synthesis*. Kluwer Academic Publishers, Boston, 1991.

[121] D. E. Thomas, E. D. Lagnese, R. A. Walker, J. A. Nestor, J. V. Rajan, and R. L. Blackburn. *Algorithmic and Register-Transfer Level Synthesis: The System Architect's Workbench*. Kluwer Academic Publishers, Boston, 1990.

[122] H. Trickey. "Flamel: A high-level hardware compiler". In *IEEE Trans. CAD*, volume 6, pages 259–269, Mar. 1987.

[123] E. Trischler. "Incomplete scan path with an automatic test generation approach". In *Proc. Int. Test Conf.*, pages 153–162, Philadelphia, Nov. 1980.

[124] F.-S. Tsai and Y.-C. Hsu. "STAR: A system for hardware allocation in data path synthesis". In *IEEE Trans. CAD*, pages 1053–1064, Sep. 1992.

[125] C.-J. Tseng and D. P. Siewiorek. "FACET: A procedure for automated synthesis of digital systems". In *Proc. Design Automation Conf.*, pages 490–496, Miami Beach, June 1983.

[126] C.-J. Tseng and D. P. Siewiorek. "Automated synthesis of data paths in digital systems". In *IEEE Trans. CAD*, volume 5, pages 379–395, July 1986.

[127] C.-J. Tseng, R.-S. Wei, S. G. Rothweiler, M. M. Tong, and A. K. Bose. "Bridge: A versatile behavioral synthesis system". In *Proc. Design Automation Conf.*, pages 415–420, Anaheim, CA, June 1988.

[128] J. S. Turner. "Almost all k-colorable graphs are easy to color". In *Journal of Algorithms*, volume 9, pages 63–82, 1988.

[129] J. Vanhoof, K. V. Rompaey, I. Bolsens, G. Goossens, and H. D. Man. *High-Level Synthesis for Real-Time Digital Signal Processing*. Kluwer Academic Publishers, Boston, 1993.

[130] P. Vishakantaiah, J. A. Abraham, and M. S. Abadir. "Automatic test knowledge extraction from VHDL (ATKET)". In *Proc. Design Automation Conf.*, pages 273–287, Anaheim, CA, June 1992.

[131] P. Vishakantaiah, T. Thomas, J. A. Abraham, and M. S. Abadir. "AMBIANT: Automatic generation of behavioral modifications for testability". In *Proc. Int. Conf. Computer Design*, pages 63–66, Boston, Oct. 1993.

[132] K. Wakabayashi. "Cyber: High level synthesis system from software into ASIC". In R. Camposano and W. H. Wolf, editors, *High-Level VLSI Synthesis*. Kluwer Academic Publishers, Boston, 1991.

[133] L.-T. Wang and E. J. McCluskey. "Concurrent built-in logic block observer (CBILBO)". In *Proc. Int. Symp. Circuits & Systems*, pages 1054–1057, San Jose, CA, May 1986.

[134] M. J. Y. Williams and J. B. Angell. "Enhancing testability of large scale integrated circuits via test points and additional logic". In *IEEE Trans. Computers*, volume C-22, pages 46–60, Jan. 1973.

[135] N.-S. Woo. "A global, dynamic register allocation and binding for a data path synthesis system". In *Proc. Design Automation Conf.*, pages 505–510, Orlando, June 1990.

Index

ADEPT, 186–188, 190
ADPS, 47
allocation, 19
    basic concept, 38
    bipartite matching, 45
    branch-and-bound, 46
    clique partitioning, 42
    global approach, 41
    integer linear programming (ILP), 47
    interactive/constructive approach, 41
    left edge algorithm (LEA), 41
AMBIANT, 193
arithmetic built-in self test (ABIST), 196
ARTEST, 190
assignment loop, 177
ATKET, 193
automatic test pattern generation (ATPG), 4

basic block, 19, 124, 125

BBLEA, 65, 67, 69–71, 95, 96, 137–139
BBREA, 67, 69–71, 95, 96, 137–139
behavioral modeling, 19–21
BETA, 190, 192
BETS, 168, 169, 176–181
BILBO, 52–54, 152, 153, 155, 158, 162, 196, 197
bipartite-matching, 45
birth time, 36
BIST, 6, 7, 52–54, 151
BLIF, 48, 75, 97, 142
bottom layer, 128
boundary variable, 20, 89, 90, 94–96, 101, 141, 148
branch-and-bound, 46, 64, 65, 68–70, 95, 96, 137, 138
branch-and-bound left edge algorithm, *see* BBLEA
branch-and-bound right edge algorithm, *see* BBREA
Bridge, 123
built-in logic block observation

(BILBO) register, *see* BILBO
built-in self test, *see* BIST

CALLAS, 2
CAMAD, 23
Cathedral, 2, 25, 176
CBILBO, 53, 152, 158, 160, 162, 164, 196, 197
CDFG, 124–126, 176, 178–181, 187, 190, 191
CDFG loop, 177–179, 181
channel routing, 42
CHARM, 41
Chen-Karnik-Saab method, 190
child layer, 127, 128, 137
circuit testing, 3
clique partitioning, 31, 39, 42, 158, 175
comparability graph, 74
compatibility graph, 42, 158, 175, 176
concurrent BILBO (CBILBO) register, *see* CBILBO
conditional branch, 29, 123–126, 128, 129, 136, 141, 142
conditional resource sharing, 123
conflict graph, 158
conflict probability, 166
control flow graph (CFG), 29
control step, 22, 31, 33

control-data flow graph, *see* CDFG
controllability, 6
coverage probability, 166
critical path, 22, 27
cut, 33
Cyber, 123

data flow graph, *see* DFG
data path circuit graph, *see* DPCG
death time, 36
dedicated register file, 176
deflection operators, 180
density of encoding, 6
design for test, *see* DFT
DFG, 12, 19–22, 24–28, 35, 36, 38, 93, 107, 112–114, 116, 124, 125
  acyclic, 20
  cyclic, 20
  cyclic data flow, 20
  data flow, 20
DFT, 14, 16, 58, 83, 97, 103–105, 142, 150
distribution graph, 27
DPCG, 50, 51, 61–63, 72, 73, 79–81, 84–93, 103, 104, 110, 130, 131, 136, 177

EFG, 187, 188, 217

INDEX

ELF, 25, 41
EMUCS, 41
end block, 126
entropy, 154, 155
error, 3
execution flow graph, *see* EFG
extended timed petri net (ETP) model, 193
EXU S-graph, 194

F-path, 168, 172
FACET, 24, 42
false block, 125
fault
    single stuck-at model, 3
fault coverage, 3
fault efficiency, 3
FDS, 116–120, 122
Flamel, 24
force, 25, 27
force-directed scheduling, *see* scheduling, FDS
freedom, 26

Genesis, 168–172, 175, 176, 195
graph coloring, 158, 162
greedy left edge algorithm, *see* LEA

HAL, 25, 27, 99, 104

hard conflict, 164, 165
HCDFG, 124, 127–130, 133–137, 141–144, 146, 148–150
head block, 126
hierarchical control-data flow graph, *see* HCDFG
hierarchical testing, 169, 170, 195
high-level synthesis, 50
    area metric, 13
    delay metric, 13
    design space, 8, 14
    flow, 48
    in general, 2, 7, 12
high-level test synthesis, *see* HTS
HIS, 2
HITEC, 190
hot potato transformation, 180
HTS, 83, 93, 94, 96, 97, 99, 105, 123, 127, 130, 135, 136
    BIST, 52–54, 151–167
    brief review, 52–55
    in general, 13, 14
    RTL, 186–192
    scan, 54–55, 168–186
Hwangs method, 123
HYPER, 177

I-path, 168, 172, 196

input register, 38, 64, 67, 93, 95, 96, 144, 147
interconnection allocation, 39, 41, 42, 48, 58, 60, 61, 64, 94, 136
interval graph, 33
IO register, 38, 88, 89, 92, 95, 133
iterative array model, 5

Kims method, 123

layer tree, 127, 128, 137, 139, 140
LEA, 42, 43, 58, 59, 73, 78, 145, 147, 150
Lee-Reddy algorithm, 97, 103, 105, 142, 148
Liapunov stability theorem, 156
lifetime analysis, 37
lifetime of a variable, 36
lifetime table, 37
logic synthesis, 7, 13
loop construct, 20, 83, 88, 92
LYRA & ARYL, 45, 99, 104

macro testing, 195
    test protocol, 195
MAHA, 26, 123
maximal BIST strategy, 154, 156
MIMOLA, 24, 47
MISR, 52, 152, 154, 156, 164–167

mobility, 21, 25, 114, 115
mobility path scheduling, *see* MPS
module allocation, 27, 38, 39, 41, 42, 47, 48, 60–62, 64, 69, 73, 74, 85, 92, 94, 102, 130, 136, 138, 158, 175, 180
module allocation graph (MAG), 38
module sharing preference, 39
Monte Carlo simulation, 155
Move Frame Scheduling (MFS), 156
MPS, 112–114, 116–121, 218
multiple-input signature register (MISR), *see* MISR

network simplex method, 183
non-scan, 83, 93, 94, 97, 100–102, 105, 116, 124, 136, 142, 148, 181
non-self-loop, 51, 52, 186

observability, 6
OPUS, 180, 190
output register, 38, 62, 64, 67, 79, 92, 93, 95, 96, 138
overlapping sub-interval, 33

parent layer, 127, 128, 137, 139
partial scan, 83, 94, 97, 102, 103, 105, 124, 136, 141, 142, 148
partial-intrusion BIST strategy, 164

INDEX

219

PHITS, 2, 16, 17, 48, 49, 57, 64, 67, 74, 75, 78, 79, 81, 83, 97, 102, 105, 107, 112, 113, 116, 121, 124, 142, 148
PHITS-NS, 97, 99–105, 116, 117, 119, 120, 142, 144, 145, 147, 150
PHITS-PS, 99, 101–105, 142, 148, 150, 168
Princeton HI-level Test Synthesis system, *see* PHITS, PHITS-NS, PHITS-PS

RALLOC, 152, 158, 162, 163, 197
random test pattern generation (RTPG), *see* RTPG
randomness, 154–156
REAL, 41
register allocation, 38, 41–43, 45–48, 58–60, 62, 64, 69, 70, 72, 73, 81, 85, 92, 94–96, 100–102, 129, 130, 136–141, 158, 160, 162, 163, 175, 181
register conflict graph, 158, 160–163
  conflict edge, 158
  self-adjacent constraint edge, 160
register file clique, 177, 178
register-transfer level (RTL)
  architecture, 2, 12, 13

synthesis, 7, 13
  test synthesis, 186–192
regular module library, 38
RTPG, 52, 152, 154, 156, 164, 165, 167

S-graph, 50, 194
scan variable, 95, 96, 101, 178, 179
scheduled data flow graph, *see* SDFG
scheduling, 19
  ALAP scheduling, 24
  ASAP scheduling, 24
  basic concept, 21
  critical path scheduling, 26
  force-directed scheduling, 27
  integer linear programming (ILP) scheduling, 35
  iterative/constructive approach, 23
  list scheduling, 25
  path-based scheduling, 29
  transformational approach, 23
SDFG, 22, 36–42, 57, 58, 60, 61, 64, 65, 70, 72, 75–77, 79, 81, 84–86, 88–90, 94–99, 101, 102, 108, 110, 116, 117, 119
Sehwa, 123
select BIST strategy, 155, 156, 164
self-adjacent register, 52, 152–154,

158, 162, 196, 197
self-loop, 51, 53–55, 154, 178, 186, 196
sequential depth, 6, 50, 57, 60, 62–64, 70–73, 79, 92, 93, 131–133, 135–138
sequential false loop, 177, 193
sequential loop, 6, 52, 83–95, 105, 130–133, 135, 136, 140, 141, 186
sequential path, 50, 51, 60, 85, 86, 88, 90, 92–95, 130–132, 135, 138, 140
serial scan in/out, 52, 152
shortest path algorithm, 74
signature, 52, 152
simple schedule, 22
SIS, 48, 75, 97, 116, 142
slack, 21, 113–115
soft conflict, 164, 165
Splicer, 25, 46, 99, 104
SR1, 60, 61, 64, 67, 83, 93, 107–110
SR2, 63, 64, 67, 70, 71, 83, 93, 107, 110, 111
SR3, 93, 107, 110
SR4, 110, 112, 114, 115
STAR, 47
STEED, 48, 63, 72, 73, 75, 79, 88, 97, 116, 117, 132, 142

subspace state coverage, 196
SYNCBIST, 151, 152, 164, 166, 167
SYNTEST, 54, 152, 156, 157, 164, 197

TBINET, 55, 168, 169, 181, 185
test environment, 172, 173, 175, 176, 195
test path, 164–167
test point, 168
test sequence range, *see* TSR
testability analysis, 50
testable functional block (TFB), 153
transparency, 155, 156
true block, 125
TSR, 187, 220

unconditional resource sharing, 123
urgency, 25

Yorktown Silicon Compiler, 23

zone scheduling, 36